Organic Rankine Cycles for Waste Heat Recovery - Analysis and Applications

Edited by Silvia Lasala

Published in London, United Kingdom

IntechOpen

Supporting open minds since 2005

Organic Rankine Cycles for Waste Heat Recovery - Analysis and Applications
http://dx.doi.org/10.5772/intechopen.77463
Edited by Silvia Lasala

Contributors
Silvia Lasala, Andrés Pina-Martinez, Jean-Noel Jaubert, Christophe Coquelet, Alain Valtz, Pascal Theveneau, Qiang Liu, Tzu-Chen Hung, Yongqiang Feng, Syamimi Saadon, Salmah Md Saiful Islam

First published in London, United Kingdom, 2020 by IntechOpen
IntechOpen is the global imprint of INTECHOPEN LIMITED, registered in England and Wales, registration number: 11086078, 7th floor, 10 Lower Thames Street, London, EC3R 6AF, United Kingdom
Printed in Croatia

British Library Cataloguing-in-Publication Data
A catalogue record for this book is available from the British Library

Additional hard and PDF copies can be obtained from orders@intechopen.com

Organic Rankine Cycles for Waste Heat Recovery - Analysis and Applications
Edited by Silvia Lasala
p. cm.
Print ISBN 978-1-78985-473-2
Online ISBN 978-1-78985-474-9
eBook (PDF) ISBN 978-1-83880-376-6

For EU product safety concerns:
IN TECH d.o.o., Prolaz Marije Krucifikse Kozulić 3, 51000 Rijeka, Croatia, info@intechopen.com or visit our website at intechopen.com.

Meet the editor

Silvia Lasala is an assistant professor at the National School of Chemical Engineering (ENSIC) of the University of Lorraine. She currently provides lectures in chemical and energy engineering courses and deals with the development of various research topics at the Laboratory Reaction and Process Engineering (LRGP) located in Nancy (France). The main purpose of her research is to contribute to the improvement of the efficiency of energy conversion technologies for electrical and thermal applications. In particular, she is working on the development of a breakthrough energy conversion technology based on the use of high-energy-density fluids for closed-power cycles. Another research area of expertise is the quantification and improvement of the accuracy of thermodynamic models used for the description of fluid properties. Further to the publication of different papers and book chapters, she is also a co-editor of the International Journal of Energy, Environment and Economics.

Contents

Preface

In the landscape of today's energy conversion technologies, Organic Rankine Cycles (ORC) play a key role in the increase of the efficiency of energy-intensive industrial processes. Indeed, the adoption of ORC enables for the recovery of the available waste heat and its conversion to useful power. In particular, this technology is currently the most investigated one to convert the largest amount of available - and otherwise lost - energy: low-temperature waste heat. Many applications of ORC exist within the framework of waste heat recovery and this book aims to present some noteworthy examples of studied and available low- and high-power applications.

The crucial difference between ORCs lies in the working fluid they employ. The use of a specific working fluid imposes specific thermodynamic properties and, thus, establishes cycle layout and types of components. The choice of the working fluid mainly depends on the thermal characteristics of the available thermal source (temperature profile and heat grade) and should ideally result from an optimisation process. The current limit of applied optimisation processes mainly lies in the scarce flexibility and low accuracy of applied thermodynamic models and in the declared lack of experimental data to validate calculations. The first section of this book aims to provide researchers with proper tools to calculate and validate thermophysical properties of ORC working fluids. This section contains two chapters. The first one presents an accurate and predictive thermodynamic model to enable the reliable calculation of thermodynamic properties of thousands of pure fluids and their mixtures. The second chapter introduces a complete set of experimental techniques for the measurement of thermophysical properties of ORC fluids. The second section of the book introduces some theoretical and experimental studies of ORCs: a review of different supercritical ORC (Chapter 3), ORC for waste heat recovery from fossil-fired power plants (Chapter 4), and an experimental detailed characterization of a small-scale ORC of 3 kW operating with either pure fluids or mixtures (Chapter 5).

I wish to thank all the authors who contributed to the realisation of this book and the members of the IntechOpen publishing process staff for their fruitful and efficient cooperation.

Silvia Lasala
University of Lorraine,
France

Section 1

Working Fluid Characterization

Chapter 1

A Predictive Equation of State to Perform an Extending Screening of Working Fluids for Power and Refrigeration Cycles

Silvia Lasala, Andrés-Piña Martinez and Jean-Noël Jaubert

Abstract

This chapter presents the features of the *Enhanced-Predictive*-PR78 equation of state (E-PPR78), a model highly suitable to perform "physical fluid screening" in power and refrigeration cycles. It enables, in fact, the accurate and predictive (i.e., without the need for its preliminary optimization by the user) determination of the thermodynamic properties of pure and multicomponent fluids usable in power and refrigeration cycles: hydrocarbons (alkanes, alkenes, alkynes, cycloalkane, naphthenic compounds, and so on), permanent gases (such as CO_2, N_2, H_2, He, Ar, O_2, NH_3, NO_2/N_2O_4, and so on), mercaptans, fluorocompounds, and water. The E-PPR78 equation of state is a developed form of the Peng-Robinson equation of state, which enables both the predictive determination of binary interaction parameters and the accurate calculation of pure fluid and mixture thermodynamic properties (saturation properties, enthalpies, heat capacities, volumes, and so on).

Keywords: thermodynamic cycle, pure working fluid, mixture, thermodynamic models, translated-E-PPR78

1. Introduction

Performance and design of closed power and refrigeration cycles are basically driven by the thermodynamic properties of their working fluids. This is the reason why, since the early 1900s, many researchers have been stressing over the importance of optimizing the working fluid of these cycles and of selecting a proper thermodynamic model to accurately calculate their properties.

Two approaches are currently applied to seek the optimal working fluid. The first strategy consists in considering a limited number of existing pure fluids, the "physical fluid screening." Alternatively, authors apply a *product design* approach, consisting in considering the molecular parameters of the working fluid as optimization variables; the resulting optimal fluid is thus fictive and is named here "fictive fluid screening."

The application of the "physical fluid screening" is preferably associated with the use of equations of state whose accuracy has been properly validated over experimental data of the considered set of existing fluids (see, e.g., [1–5]). The preferred modeling option lies in the use of multi-parameter equations of state such as Helmholtz energy-based equations of state optimized by NIST (e.g., the GERG [6], the Span and Wagner [7], and so on), m-Benedict-Webb-Rubin (BWR) [8], Bender [9], and so on. Despite being highly accurate, these equations of state require the availability of a huge number of fluid-specific parameters, and their optimal values are thus provided by the model developer. An interesting chapter [10] has been recently published by Bell and Lemmon to spread the use of multi-parameter equations of state in the ORC community. However, at their current state of development, these models are thus not sufficiently flexible to be used in a screening approach extended to a population of hundreds of existing pure fluids and mixtures. For the same reason, the use of these multi-parameter models in a "fictive fluid screening" approach is inappropriate. To provide the reader with an order of magnitude, more than 1000 of fluids could be considered in this physical fluid screening procedure. The Design Institute for Physical Properties (DIPPR) currently provides accurate experimental data in a database (DIPPR 801) for 2330 pure fluids. Refprop 10.0 (NIST) [11] currently allows for the accurate representation of only 147 fluids.

To extend the range of considered fluids, studies present in the literature also consider the use of more flexible equations of state, that is, models characterized by a low number of parameters. If we focus on studies about closed power cycles, the equations of state, which have mainly been applied, are as follows: PC-SAFT-based model [12, 13] (which requires three molecule-specific parameters) in [14–16], BACKONE equation of state [17] (with four molecule-specific parameters) in [18], and the standard Peng-Robinson equations of state [19, 20] (with three parameters for each pure fluid) in [21–23]. These authors considered a different number of fluids. The one counting the highest considered number of fluids is the study by Drescher and Brüggemann [21], with 700 pure fluids. To our knowledge, all the other studies count less than 100 fluids (generally between 10 and 30). Peng-Robinson equation of state is currently the most flexible model to perform an extensive "physical fluid screening" of power and refrigeration working fluids. One of the main conclusions of authors who applied and compared different thermodynamic models (which is, unfortunately, rarely the case—we just found one study) is that the use of the Peng-Robinson equation of state is reliable in comparison with more accurate—but less flexible—multi-parameter equations of state [23].

Since 2004, Jaubert and co-workers have started publishing an improved version of the Peng-Robinson equation of state (version of the year 1978, PR78), the "Enhanced-Predictive-Peng-Robinson-78" (*E*-PPR78) [24–40]. Differently from PR78, this model is entirely able to **predict** the properties of mixtures *without the need for its preliminary calibration* over experimental data; moreover, the adjective *enhanced* has been juxtaposed to its previous name (PPR78) in 2011 [41] to highlight the **improved accuracy** in calculating **mixing enthalpies and heat capacities** (with respect to the original PPR78 model).

This model is widely used in the Chemical Engineering community but, inexplicably, remains unknown in the Energy Engineering one. The aims of this chapter are thus to present this model, to outline the proper way to apply it according to the latest advancements over pure fluid modeling [42–46], and to perform the screening of pure and/or multicomponent working fluids for power and refrigeration cycles.

2. From Peng-Robinson to E-PPR78 equation of state

The E-PPR78 model is an improved version of the equation of state published in 1978 by Peng and Robinson, the PR78 equation of state. This model has been developed to allow for the accurate and predictive (i.e., without the need for its optimization over experimental data) application of the Peng-Robinson equation of state to multi-component mixtures. We thus start with introducing the forerunner PR78 equation of state:

$$P = \frac{RT}{v-b} - \frac{a}{v(v+b)+b(v\text{-}b)} \tag{1}$$

When applied to the ith pure component, a in Eq. (1) corresponds to the pure-component cohesive parameter, a_i, and b to its co-volume, b_i. We will refer to a and b to indicate the mixture cohesive and co-volume parameters. We will detail in the following section how to calculate pure fluid a_i and b_i (Section 2.1) and mixture a and b (Section 2.2). Before continuing, it is worth warning the reader of the fact that the original E-PPR78 model degenerates into the standard PR78 equation of state when considering pure fluids.

2.1 PR78: the application to pure fluids

When applied to pure fluids, the standard Peng-Robinson equation of state requires the definition of parameters a_i and b_i, calculated as reported in the following:

$$\begin{cases} R = 8.314472\, J\ mol^{-1}K^{-1} \\ X = \left[1 + \sqrt[3]{4-2\sqrt{2}} + \sqrt[3]{4+2\sqrt{2}}\right]^{-1} \sim 0.253076587 \\ b_i = \Omega_b \dfrac{RT_{c,i}}{P_{c,i}} \text{ with: } \Omega_b = \dfrac{X}{X+3} \sim 0.07780 \\ a_i(T) = a_{c,i}\alpha_i(T) \text{ with } \begin{cases} a_{c,i} = \Omega_a \dfrac{R^2T_{c,i}^2}{P_{c,i}} \text{ and } \Omega_a = \dfrac{8(5X+1)}{49-37X} \sim 0.457235529 \\ \alpha_i(T) \text{ is the so-called } \alpha - \text{function} \end{cases} \end{cases} \tag{2}$$

The standard Peng-Robinson equation of state incorporates the Soave α-function [19, 47]:

$$\begin{cases} \alpha_i(T) = \left[1 + m_i\left(1 - \sqrt{\frac{T}{T_{c,i}}}\right)\right]^2 \\ \text{if } \omega_i \le 0.491 \text{ then } m_i = 0.37464 + 1.54226\omega_i - 0.26992\omega_i^2 \\ \text{if } \omega_i > 0.491 \text{ then } m_i = 0.379642 + 1.48503\omega_i - 0.164423\omega_i^2 + 0.016666\omega_i^3 \end{cases} \tag{3}$$

However, in the last 4 years, two improved (i.e., thermodynamically consistent [42, 48] and very accurate) α-functions have been developed and published [44, 46]: a fluid-specific α-function and a generalized one, respectively, based on the model Twu91 [49] and Twu88 [50]. The application of the fluid-specific α-function Twu91 optimized in [46] guarantees the highest accuracy and requires three parameters (L, M, and N) for each pure fluid (reported by Pina-Martinez et al. [46] for 1721 molecules):

$$\alpha_i(T) = \left(\frac{T}{T_{c,i}}\right)^{N_i(M_i-1)} \cdot \exp\left[L_i\left(1 - \left(\frac{T}{T_{c,i}}\right)^{M_iN_i}\right)\right]^2 \tag{4}$$

The generalized version of Twu88 [46] requires, similar to the Soave α-function, the knowledge of the acentric factor of each pure fluid and takes the following form:

$$\begin{cases} \alpha_i(T) = \left(\frac{T}{T_{c,i}}\right)^{2(M_i-1)} \cdot \exp\left[L_i\left(1-\left(\frac{T}{T_{c,i}}\right)^{2M_i}\right)\right]^2 \\ L_i(\omega_i) = 0.0925\omega_i^2 + 0.6693\omega_i + 0.0728 \\ M_i(\omega_i) = 0.1695\omega_i^2 - 0.2258\omega_i + 0.8788 \end{cases} \quad (5)$$

The alternative use of the three α-functions recalled above leads to different accuracies in the calculation of thermodynamic properties. A comparison is reported in **Table 1** between the PR equation of state incorporating the three different α-functions and pseudo-experimental data made available by DIPPR [51]. Piña-Martinez et al. also showed [46] that the modification of the α-function affects in a very negligible way the accuracy on volume calculations. To improve volumes, a further modification is required, as explained in Section 2.3.

2.2 From PR78 to E-PPR78: the application to mixtures

The application of the PR equation of state to a mixture requires the selection of mixing rules for calculating mixture cohesive and co-volume parameters, a and b. Classical Van der Waals one-fluid mixing rules are used in the original PR78 model:

$$\begin{cases} a(T,\mathbf{z}) = \sum_{i=1}^{N}\sum_{j=1}^{N} z_i z_j \sqrt{a_i(T)a_j(T)}\left(1-k_{ij}\right) \\ b(\mathbf{z}) = \sum_{i=1}^{N} z_i b_i \end{cases} \quad (6)$$

The k_{ij} parameter is the so-called binary interaction parameter characterizing the molecular interactions between molecules i and j. The most accurate application of the original PR78 model requires the empirical optimization of the k_{ij} parameter over, at least, vapor-liquid equilibrium experimental data.

In 2004, Jaubert and Mutelet [24] proposed a model to *predictively* calculate the k_{ij} parameter by means of the application of a group contribution method. This method allows to estimate and predict the k_{ij} parameter by combining the molecular characteristics of elementary groups in which each molecule can be subdivided. This model is the most physically grounded model to determine the k_{ij} binary

α-Function	Soave, Eq. (3)	Generalized Twu88, Eq. (5)	Fluid-specific Twu91, Eq. (4)
MAPE on P^{sat} (1721 compounds)	2.8%	1.8%	1.0%
MAPE on $\Delta_{vap}H$ (1453 compounds)	3.1%	2.7%	2.9%
MAPE on $c_p^{sat,\ liquid}$ (829 compounds)	7.1%	4.1%	2.0%

Data have been collected from [46].

Table 1.
Comparison of the mean average percentage errors (MAPEs) calculated with PR incorporating either the Soave α-function or the generalized Twu88 α-function or fluid-specific Twu91 α-function.

interaction parameters of PR-78 equation of state, and its use is extremely recommended to predictively calculate thermodynamic properties of multi-component mixtures. The expression provided by this model to predictively calculate the binary interaction parameter is as follows:

$$k_{ij}(T) = \frac{-\frac{1}{2}\left[\sum_{k=1}^{N_g}\sum_{l=1}^{N_g}\left(\alpha_{ik}-\alpha_{jk}\right)\left(\alpha_{il}-\alpha_{jl}\right)A_{kl}\left(\frac{298.15}{T/K}\right)^{\left(\frac{B_{kl}}{A_{kl}}-1\right)}\right]-\left(\frac{\sqrt{a_i(T)}}{b_i}-\frac{\sqrt{a_j(T)}}{b_j}\right)^2}{2\frac{\sqrt{a_i(T)a_j(T)}}{b_ib_j}} \tag{7}$$

where a_i and b_i are the energy and co-volume parameters of the ith molecule, given in Eq. (2); N_g is the number of different groups defined by the method; and α_{ik} is the fraction of molecule i occupied by group k (occurrence of group k in molecule i divided by the total number of groups present in molecule i). A_{kl} and B_{kl}, the group-interaction parameters, are symmetric, $A_{kl} = A_{lk}$ and $B_{kl} = B_{lk}$ (where k and l are two different groups), and empirically determined by correlating experimental data. Also, $A_{kk} = B_{kk} = 0$. The inclusion of this predictive expression for k_{ij} in the PR78 equation of state results in the **Predictive-PR78 (PPR78)**.

It is worth recalling the historical development of the process of optimization of A_{kl} and B_{kl} provided by the model developers. These parameters have initially been optimized over only vapor-liquid equilibrium data of binary mixtures. The model resulting from the use of these so-optimized group contribution parameters is called PPR78 (Predictive-Peng-Robinson equation of state). Lately, authors recognized that the inclusion of enthalpy and heat capacity data in the optimization process does not affect the accuracy in modeling VLE properties but improves extraordinarily the accuracy in calculating enthalpies and heat capacities of mixtures. So, starting from the year 2011 [41], published A_{kl} and B_{kl} have been obtained by minimizing the errors between model calculations and experimental data relative to VLE, mixing enthalpy and heat capacity properties. The model resulting from the inclusion of these group contribution parameters is called **Enhanced-Predictive-PR78 equation of state (E-PPR78)**. The last optimized values of A_{kl} and of B_{kl} are reported in Table S1 of Supplementary Material of [39] for 40 molecular groups.

The optimization of these parameters has been performed over more than 150,000 experimental data and developed over more than 15 years. Even if preferable, that would be quite time-expensive if there was the need to re-optimize these group contribution parameters when changing any feature of the cubic equation of state (e.g., the α-function) or the cubic equation of state itself. Thankfully, it has been demonstrated [52] that it is possible to rigorously determine k_{ij} of any equation of state, knowing those of the original E-PPR78. In particular, it is possible to easily replace the Soave α-function, originally present in E-PPR78, with one of the improved functions presented in Section 2.1 (Eqs. **(4)** and **(5)**) and to use A_{kl} and B_{kl} parameters of the Soave-based E-PPR78 by applying, instead of Eq. (7):

$$k_{ij}(T) = \frac{-\frac{1}{2}\left[\sum_{k=1}^{N_g}\sum_{l=1}^{N_g}\left(\alpha_{ik}-\alpha_{jk}\right)\left(\alpha_{il}-\alpha_{jl}\right)A_{kl}\left(\frac{298.15}{T/K}\right)^{\left(\frac{B_{kl}}{A_{kl}}-1\right)}\right]-\left(\frac{\sqrt{a_i^{\mathrm{mod}}(T)}}{b_i}-\frac{\sqrt{a_j^{\mathrm{mod}}(T)}}{b_j}\right)^2}{2\frac{\sqrt{a_i^{\mathrm{mod}}(T)a_j^{\mathrm{mod}}(T)}}{b_ib_j}} \tag{8}$$

With respect to Eq. (7), this expression incorporates the pure component energy parameters calculated from the modified α-function. If we consider Twu α-function, we will thus use a_i^{mod} given by:

$$a_i^{\text{mod}}(T) = a_{c,i}\alpha_i^{\text{mod}}(T) \tag{9}$$

It is worth observing that for the systems for which the Soave α-function is already very accurate (i.e., *mean average percentage errors* of the order of 1% for saturation pressures and of 2% for vaporization enthalpies and liquid heat capacities), the k_{ij} in Eq. (7) (i.e., the standard *E*-PPR78 model, with the Soave function) is able to provide the best reproduction of mixture data. The alternative use of a more accurate *α-function* (which thus improves pure fluid calculations) and Eq. (8), to enable the use of original A_{kl} and B_{kl} group contribution parameters optimized with the original Soave-based *E*-PPR78, slightly deteriorates the results on mixtures (e.g., in the case of mixtures of alkanes). Clearly, the best would consist in re-optimizing all group contribution parameters using the best α-function directly in Eq. (7) instead of using the less-time-consuming Eq. (8) to derive the modified k_{ij}(T) parameters. However, even adopting the simplified approach consisting in using Eq. (8), the predictive capability of this model remains very accurate for modeling

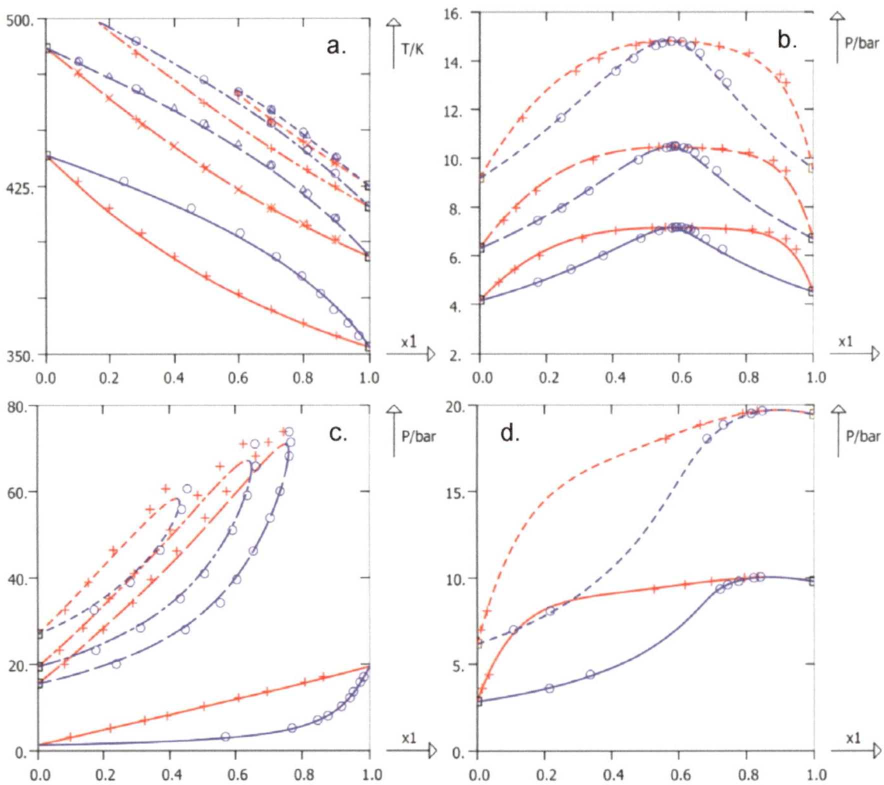

Figure 1.
Isobar vapor-liquid equilibrium phase diagrams for the system n-butane (1)–n-hexane (2) (a) and isothermal vapor-liquid equilibrium phase diagrams for the system 1-butene (1)-R610 (2) (b), CO_2 (1)-R134a (2) (c), R116 (1)-ethylene (2) (d). Lines represent calculations with E-PPR78 with Twu91 alpha-function, Eq. (4) (b). Bubble points are indicated in red, dew points in blue. Black points represent calculated pure component saturation pressures. (a) P *(bar) = 10.132 (continuous line), 25.855 (long-dashed line), 32.75 (long- and short-dashed line), and 37.921 (short-dashed line); (b)* T *(K) = 312.92 (continuous line), 327.93 (long-dashed line), and 342.93 (short-dashed line); (c)* T *(K) = 252.95 (continuous line), 329.60 (long-dashed line), 339.10 (long- and short-dashed line), and 354.00 (short-dashed line); and (d)* T *(K) = 251.00 (continuous line) and 275.00 (short-dashed line).*

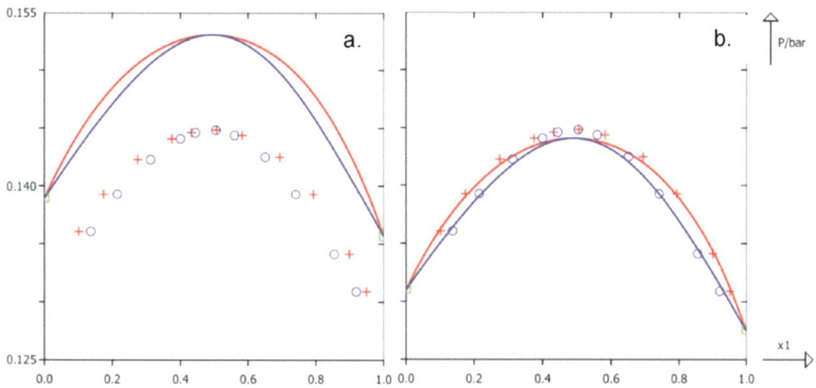

Figure 2.
Isothermal VLE diagrams of the benzene (1)–cyclohexane (2) system, at 298.15 K. Lines represent calculations with standard E-PPR78 (a) and E-PPR78 with Twu91 alpha-function, Eq. (4) (b). Bubble points are indicated in red, dew points in blue. Green points represent calculated pure components saturation pressures.

closed power cycle working fluids and refrigerants (see the examples reported in **Figure 1**).

However, there are systems for which the Soave model is very inaccurate and the use of Twu α-function with Soave-based E-PPR78 A_{kl} and B_{kl} parameters highly improves results. By way of example, we present a pivotal binary mixture, benzene-cyclohexane, for which the standard PR equation of state (i.e., with the Soave α-function) does predict in a very inaccurate way of pure component saturation pressures. The original *E*-PPR78 equation of state, based on standard PR, is thus not very accurate in predictively modeling mixture saturation pressures because of the basic incapacity of the PR equation of state in modeling pure fluid properties (see in **Figure 2a**). However, if the Soave α-function is replaced with a more accurate α-function (given, e.g., by Eq. (4)) and if we then use Eq. (8) (with the Soave-based E-PPR78 A_{kl} and B_{kl} parameters reported in [39]) to represent benzene (formed by six groups CH_{aro}) and cyclohexane (formed by six groups $CH_{2,cyclic}$), we obtain the graph as shown in **Figure 2b**. The accuracy is thus strongly improved without the need of re-optimizing any parameter.

Considering the above remarks, we suggest the replacement of the Soave α-function with the Twu one, in E-PPR78, thus applying Eq. (8) and Soave-based E-PPR78 group contribution A_{kl} and B_{kl} parameters.

2.3 Volume correction

It is well known that one of the main limitations of cubic equations of state is their inaccuracy in high predicting liquid densities. Péneloux et al. [53] showed that it was possible to come up with this problem by adding a *translation term to the volume*. This translation consists in correcting the volume resulting from the resolution of the cubic equation of state (Eq. (1)) as follows:

- In case of pure fluids:

$$v_i^t(T,P) = v_i(T,P) - c_i \tag{10}$$

- In case of mixtures:

$$\begin{cases} v^t(T,P,z) = v(T,P,z) - c \\ c = \sum_i^{N_c} c_i \cdot z_i \quad \text{(linear mixing rule for } c\text{)} \end{cases} \tag{11}$$

In a recent publication, some accurate generalized (i.e., predictive) expressions for the translation term are optimized over 475 compounds, available in the DIPPR. For the Peng-Robinson equation of state, it is provided as follows:

$$c_i = \frac{RT_{c,i}}{P_{c,i}}(0.1975 - 0.7325 \cdot z_{RA,i}) \tag{12}$$

A databank of Rackett compressibility factors, z_{RA}, for 1489 components is available in Supplementary Material of [46]. The application of this translation has been observed to greatly improve the mean average percentage errors on calculated volumes. Considering the same 1489 pure fluids, the authors attested, in the same work, that the error in calculating the volume of the liquid phase at saturation condition is reduced from 8.7% (PR without translation) to 2.2% (PR with translation in Eq. (12)). If z_{RA} is not available, authors suggested the use of the following expression, where the translation term is only a function of the acentric factor.

$$c_i = \frac{RT_{c,i}}{P_{c,i}}(0.0096 + 0.0049 \cdot \omega_i) \tag{13}$$

Jaubert et al. [45] were able to demonstrate that entropy (s), internal energy (u), Helmholtz energy (a), constant pressure and constant volume heat capacity (c_p and c_v), vapor pressure (P^{sat}), and all properties change of vaporization ($\Delta_{vap}H$, $\Delta_{vap}S$, $\Delta_{vap}U$, $\Delta_{vap}A$, $\Delta_{vap}C_p$, and $\Delta_{vap}C_v$) of pure fluid properties are not influenced by a temperature-independent volume translation.

It can be thus deduced that the addition of a translation term and the modification of the α-function have unlinked effects: the utilization of a volume translation improves volume calculations without affecting the abovementioned thermodynamic properties, while the use of an improved α-function improves subcritical and supercritical properties without deteriorating density calculations (see Conclusion reported in Section 2.1).

The application of both the consistent-Twu α-function (either Eq. (4) or Eq. (5)) and the volume translation in Eq. (12) results in the most accurate generalized cubic equation of state available in the literature.

For completeness, we would like to observe that, other than volume, also enthalpy and speed of sound are affected by the inclusion of a temperature-independent volume translation term (see [45]). However, the impact of such a translation on the calculation of enthalpy differences and of speed of sound is really negligible. In fact, it can be mathematically demonstrated from the use of relations presented in [45] that the enthalpy variation calculated with the translated cubic equation of state, Δh_t, and the one calculated with the nontranslated form, Δh_0, are related by the following relation:

$$\frac{\overbrace{h_t(T_1,P_1) - h_t(T_2,P_2)}^{\Delta h_t}}{\underbrace{h_0(T_1,P_1) - h_0(T_2,P_2)}_{\Delta h_0}} = 1 - \frac{c_i \cdot (P_1 - P_2)}{\Delta h_0} \tag{14}$$

So, first, it can be observed that isobar enthalpy variations are not affected by the inclusion of a volume translation term. Moreover, it can be shown that, in general, for temperature and pressure conditions relevant for power and refrigeration cycle applications, the term $c_i \cdot (P_1 - P_2)/\Delta h_0$ is much lower than 0.001 for gaseous systems and lower than 0.005 for liquid systems. As regards the speed of sound, it can be mathematically derived that ratio between the speed of sound determined with the translated cubic equation of state, w_t, and the one calculated with the nontranslated EoS, w_0, is given by:

$$\frac{w_t(T,P)}{w_0(T,P)} = \sqrt{1 - \frac{c}{v(T,P)}} \tag{15}$$

which generally varies between 0.990 and 1.020 for liquid systems and is equal to 1.000 for vapor systems. These quantifications have been performed considering toluene, R134a, butane, propane, and ammonia.

3. Conclusion

Despite its simplicity and flexibility, E-PPR78 is a model that guarantees one of the most reliable predictive determinations of the thermodynamic properties of working fluids for power and refrigeration cycles. Being by definition a predictive model, its use is highly suggested to look for the best working fluid candidate over thousands of pure and multi-component fluids.

In this chapter, we presented the model and suggested to modify the Soave-based-original-*E*-PPR78 model by using the Twu α-function, to allow for the more precise representation of systems for which the Soave one is not sufficiently accurate. Finally, we recalled that the inclusion of a volume translation term in the E-PPR78 model highly improves the errors in the calculation of densities without affecting the rest of the, already accurate, properties.

Author details

Silvia Lasala*, Andrés-Piña Martinez and Jean-Noël Jaubert
Université de Lorraine, CNRS, LRGP, Nancy, France

*Address all correspondence to: silvia.lasala@univ-lorraine.fr

References

[1] Dai Y, Wang J, Gao L. Parametric optimization and comparative study of organic Rankine cycle (ORC) for low grade waste heat recovery. Energy Convers Manag. 2009;**50**(3):576-582

[2] Schuster A, Karellas S, Aumann R. Efficiency optimization potential in supercritical organic Rankine cycles. Energy. 2010;**35**(2):1033-1099

[3] Martelli E, Capra F, Consonni S. Numerical optimization of combined heat and power organic Rankine cycles—Part A: Design optimization. Energy. 2015;**90**(P1):310-328

[4] Astolfi M, Alfani D, Lasala S, Macchi E. Comparison between ORC and CO_2 power systems for the exploitation of low-medium temperature heat sources. Energy. 2018;**161**:1250-1261

[5] Walraven D, Laenen B, D'haeseleer W. Comparison of thermodynamic cycles for power production from low-temperature geothermal heat sources. Energy Conversion and Management. 2013;**66**: 220-233

[6] Kunz O, Klimeck R, Wagner W, Jaeschke M. The GERG-2004 Wide-Range Equation of State for Natural Gases and Other Mixtures [Internet]. 2007. pp. 1–555. Disponible sur: http://www.gerg.eu/publications/technical-monographs

[7] Span R, Wagner W. A new equation of state for carbon dioxide covering the fluid region from the triple-point temperature to 1100 K at pressures up to 800 MPa. Journal of Physical and Chemical Reference Data. 1996;**25**(6): 1509-1596

[8] Benedict M, Webb GB, Rubin LC. An empirical equation for thermodynamic properties of light hydrocarbons and their mixtures II. Mixtures of methane, ethane, propane, and n-butane. The Journal of Chemical Physics. 1942; **10**(12):747-758

[9] Bender E. The calculation of phase equilibria from a thermal equation of state (Engl. Transl.) [PhD thesis]. Bochum: Ruhr University; 1971

[10] Bell IH, Lemmon EW. Organic fluids for organic Rankine cycle systems: Classification and calculation of thermodynamic and transport properties. In: Organic Rankine Cycle (ORC) Power Systems: Technologies and Applications. Woodhead Publishing; 2016. pp. 91-119

[11] Lemmon EW, Huber ML, Bell IH, McLinden MO. NIST Standard Reference Database 23: Reference Fluid Thermodynamic and Transport Properties—REFPROP, Version 10.0. Standard Reference Data Program. Gaithersburg: National Institute of Standards and Technology; 2018

[12] Gross J, Sadowski G. Perturbed-chain SAFT: An equation of state based on a perturbation theory for chain molecules. Industrial and Engineering Chemistry Research. 2001;**40**(4): 1244-1260

[13] Papaioannou V, Lafitte T, Avendaño C, Adjiman CS, Jackson G, Müller EA, et al. Group contribution methodology based on the statistical associating fluid theory for heteronuclear molecules formed from Mie segments. Journal of Chemical Physics. 2014;**140**(5):054107

[14] Hattiangadi A. Working Fluid Design for Organic Rankine Cycle (ORC) Systems. 2013. Disponible sur: https://repository.tudelft.nl/islandora/object/uuid%3A492ce6c0-ab22-42d0-ba00-3cf3702eb873

[15] Lampe M, Stavrou M, Buecker HM, Gross J, Bardow A. Simultaneous

optimization of working fluid and process for organic Rankine cycles using PC-SAFT. Industrial and Engineering Chemistry Research. 2014;**53**(21): 8821-8830

[16] White MT, Oyewunmi OA, Chatzopoulou MA, Pantaleo AM, Haslam AJ, Markides CN. Computer-aided working-fluid design, thermodynamic optimisation and thermoeconomic assessment of ORC systems for waste-heat recovery. Energy. 2018;**161**:1181-1198

[17] Müller A, Winkelmann J, Fischer J. Backone family of equations of state: 1. Nonpolar and polar pure fluids. AIChE Journal. 1996;**42**(4):1116-1126

[18] Saleh B, Koglbauer G, Wendland M, Fischer J. Working fluids for low-temperature organic Rankine cycles. Energy. 2007;**32**(7):1210-1121

[19] Peng D-Y, Robinson DB. A new two-constant equation of state. Industrial and Engineering Chemistry Fundamentals. 1976;**15**(1):59-64

[20] Peng D-Y, Robinson DB. The Characterization of the Heptanes and Heavier Fractions for the GPA Peng-Robinson Programs. Gas Processors Association. Edmonton, Alberta, Canada: University of Alberta; 1978

[21] Drescher U, Brüggemann D. Fluid selection for the Organic Rankine Cycle (ORC) in biomass power and heat plants. Applied Thermal Engineering. 2007;**27**(1):223-228

[22] Invernizzi C, Iora P, Silva P. Bottoming micro-Rankine cycles for micro-gas turbines. Applied Thermal Engineering. 2007;**27**(1):100-110

[23] White MT, Sayma AI. Simultaneous cycle optimization and fluid selection for ORC systems accounting for the effect of the operating conditions on turbine efficiency. Frontiers in Energy Research. 2019;7:50

[24] Jaubert J-N, Mutelet F. VLE predictions with the Peng-Robinson equation of state and temperature-dependent k_{ij} calculated through a group contribution method. Fluid Phase Equilibria. 2004;**224**(2):285-304

[25] Jaubert J-N, Vitu S, Mutelet F, Corriou J-P. Extension of the PPR78 model (predictive 1978, Peng–Robinson EOS with temperature dependent kij calculated through a group contribution method) to systems containing aromatic compounds. Fluid Phase Equilibria. 2005;**237**(1):193-211

[26] Vitu S, Privat R, Jaubert J-N, Mutelet F. Predicting the phase equilibria of CO_2 + hydrocarbon systems with the PPR78 model (PR EoS and k_{ij} calculated through a group contribution method). Journal of Supercritical Fluids. 2008;**45**(1):1-26

[27] Vitu S, Jaubert J-N, Mutelet F. Extension of the PPR78 model (predictive 1978, Peng-Robinson EoS with temperature-dependent k_{ij} calculated through a group contribution method) to systems containing naphtenic compounds. Fluid Phase Equilibria. 2006;**243**(1–2):9-28

[28] Privat R, Jaubert J-N, Mutelet F. Addition of the nitrogen group to the PPR78 model (predictive 1978, Peng Robinson EOS with temperature-dependent kij calculated through a group contribution method). Industrial and Engineering Chemistry Research. 2008;**47**(6):2033-2048

[29] Privat R, Mutelet F, Jaubert J-N. Addition of the hydrogen sulfide group to the PPR78 model (predictive 1978, Peng-Robinson equation of state with temperature dependent k(ij) calculated through a group contribution method). Industrial and Engineering Chemistry Research. 2008;**47**(24):10041-10052

[30] Privat R, Jaubert J-N, Mutelet F. Addition of the sulfhydryl group (−SH) to the PPR78 model (predictive 1978, Peng-Robinson EoS with temperature-dependent k_{ij} calculated through a group contribution method). The Journal of Chemical Thermodynamics. 2008;**40**(9):1331-1341

[31] Privat R, Jaubert J-N. Addition of the sulfhydryl group (SH) to the PPR78 model: Estimation of missing group-interaction parameters for systems containing mercaptans and carbon dioxide or nitrogen or methane, from newly published data. Fluid Phase Equilibria. 2012;**334**:197-203

[32] Qian J-W, Privat R, Jaubert J-N. Predicting the phase equilibria, critical phenomena, and mixing enthalpies of binary aqueous systems containing alkanes, cycloalkanes, aromatics, alkenes, and gases (N_2, CO_2, H_2S, H_2) with the PPR78 equation of state. Industrial and Engineering Chemistry Research. 2013;**52**(46):16457-16490

[33] Qian J-W, Jaubert J-N, Privat R. Prediction of the phase behavior of alkene-containing binary systems with the PPR78 model. Fluid Phase Equilibria. 2013;**354**:212-235

[34] Qian J-W, Jaubert J-N, Privat R. Phase equilibria in hydrogen-containing binary systems modeled with the Peng-Robinson equation of state and temperature-dependent binary interaction parameters calculated through a group-contribution method. Journal of Supercritical Fluids. 2013;**75**: 58-71

[35] Qian J-W, Privat R, Jaubert J-N, Coquelet C, Ramjugernath D. Fluid-phase-equilibrium prediction of fluorocompound-containing binary systems with the predictive *E*-PPR78 model [Prévision en matière d'équilibre des phases de fluide des systèmes binaires contenant des fluorocomposés avec le modèle prédictif *E*-PPR78]. International Journal of Refrigeration. 2017;**73**:65-90

[36] Plee V, Jaubert J-N, Privat R, Arpentinier P. Extension of the *E*-PPR78 equation of state to predict fluid phase equilibria of natural gases containing carbon monoxide, helium-4 and argon. Journal of Petroleum Science and Engineering. 2015;**133**:744-770

[37] Xu X, Privat R, Jaubert J-N. Addition of the sulfur dioxide group (SO_2), the oxygen group (O_2) and the nitric oxide group (NO) to the *E*-PPR78 model. Industrial and Engineering Chemistry Research. 2015;**54**(38): 9494-9504

[38] Xu X, Lasala S, Privat R, Jaubert J-N. E-PPR78: A proper cubic EoS for modeling fluids involved in the design and operation of carbon dioxide capture and storage (CCS) processes. International Journal of Greenhouse Gas Control. 2017;**56**:126-154

[39] Xu X, Jaubert J-N, Privat R, Arpentinier P. Prediction of thermodynamic properties of alkyne-containing mixtures with the *E*-PPR78 model. Industrial and Engineering Chemistry Research. 2017;**56**(28): 8143-8157

[40] Qian JW, Privat R, Jaubert JN, Duchet-Suchaux P. Enthalpy and heat capacity changes on mixing: Fundamental aspects and prediction by means of the PPR78 cubic equation of state. Energy & Fuels. 2013;**27**(11): 7150-7178

[41] Qian J. Développement du modèle *E*-PPR78 pour prédire les équilibres de phases et les grandeurs de mélange de systèmes complexes d'intérêt pétrolier sur de larges gammes de températures et de pressions. Nancy: Institut National Polytechnique de Lorraine; 2011

[42] Le Guennec Y, Lasala S, Privat R, Jaubert J-N. A consistency test for

α-functions of cubic equations of state. Fluid Phase Equilibria. 2016;**427**:513-538

[43] Le Guennec Y, Privat R, Lasala S, Jaubert JN. On the imperative need to use a consistent α-function for the prediction of pure-compound supercritical properties with a cubic equation of state. Fluid Phase Equilibria. 2017;**445**:45-53

[44] Le Guennec Y, Privat R, Jaubert J-N. Development of the translated-consistent tc-PR and tc-RK cubic equations of state for a safe and accurate prediction of volumetric, energetic and saturation properties of pure compounds in the sub- and super-critical domains. Fluid Phase Equilibria. 2016;**429**:301-312

[45] Jaubert JN, Privat R, Le Guennec Y, Coniglio L. Note on the properties altered by application of a Péneloux-type volume translation to an equation of state. Fluid Phase Equilibria. 2016; **419**:88-95

[46] Pina-Martinez A, Le Guennec Y, Privat R, Jaubert J-N, Mathias PM. Analysis of the combinations of property data that are suitable for a safe estimation of consistent Twu α-function parameters: Updated parameter values for the translated-consistent *tc*-PR and *tc*-RK cubic equations of state. Journal of Chemical Engineering Data. 2018; **63**(10):3980-3988

[47] Soave G. Equilibrium constants from a modified Redlich-Kwong equation of state. Chemical Engineering Science. 1972;**27**(6):1197-1203

[48] Lasala S. Advanced cubic equations of state for accurate modelling of fluid mixtures. In: Application to CO_2 Capture Systems. Italy: Politecnico di Milano; 2016

[49] Twu CH, Bluck D, Cunningham JR, Coon JE. A cubic equation of state with a new alpha function and a new mixing rule. Fluid Phase Equilibria. 1991;**69**: 33-50

[50] Twu CH. A modified Redlich-Kwong equation of state for highly polar, supercritical systems. In: International Symposium on Thermodynamics in Chemical Engineering and Industry. 1988

[51] DIPPR 801 Database [Internet]. Disponible sur: https://www.aiche.org/dippr/events-products/801-database

[52] Jaubert J-N, Privat R. Relationship between the binary interaction parameters (k_{ij}) of the Peng–Robinson and those of the Soave–Redlich–Kwong equations of state: Application to the definition of the PR2SRK model. Fluid Phase Equilibria. 2010;**295**(1):26-37

[53] Péneloux A, Rauzy E, Fréze R. A consistent correction for Redlich-Kwong-Soave volumes. Fluid Phase Equilibria. 1982;**8**(1):7-23

Chapter 2

Experimental Determination of Thermophysical Properties of Working Fluids for ORC Applications

Christophe Coquelet, Alain Valtz and Pascal Théveneau

Abstract

The design and optimization of Organic Rankine Cycle (ORC) require knowledge concerning the thermophysical properties of the working fluids: pure components or mixtures. These properties are generally calculated by thermodynamic and transport property models (thermodynamic or equation of state or correlations). The parameters of these models are adjusted on accurate experimental data. The main experimental data of interest concern phase equilibrium properties (noncritical and critical data), volumetric properties (density and speed of sound), energetic properties (enthalpy, heat capacity), and transport properties (dynamic viscosity and thermal conductivity). In this chapter, some experimental techniques frequently used to obtain the experimental data are presented. Also, we will present some models frequently used to correlate the data and some results (comparison between experimental data and model predictions).

Keywords: working fluid design, experimental techniques, transport properties, thermodynamic properties, equations of state

1. Introduction

The utilization of energy available at low, average, and high temperature can be one solution to reduce the energy consumption particularly the fossil energy and to reduce emission of CO_2 in the atmosphere. Closed power cycle (Brayton cycle and ORC) proves to be the solution to convert heat sources into energy. Heat source can come from geothermal, solar, and biomass energy or from processes and energy systems. In general, the conditions of the heat source are fixed, and a temperature glide may be observed. In consequence, it is necessary to design the most suitable cycle architecture and to select the best working fluid in order to obtain the best performance and design of each component of the system. The selection of fluid requires thermodynamic models, and these models currently need experimental data in order to optimize their parameters. The knowledge of experimental techniques is very important to measure the thermophysical properties and to estimate their experimental uncertainties. The choice of the most suitable technique depends on the type of data to be determined, the range of pressure and temperature, the precision required, and the composition of the mixture if necessary.

In the literature several studies published concerning the investigation of working fluid for ORC applications exist. We will not present and described all the working fluid investigation. In 1985, Badr et al. [1] have examined around 68 pure working fluids (including natural fluid and hydrofluorocarbons) and given the main characteristic of the working fluids and their impacts on heat exchangers and turbine design. We can notice that in 1985, chlorofluorocarbon were not definitively banned. More recently, Saleh et al. [2] made a screening of 31 pure working fluids. They concluded that the thermal efficiency is better if the critical temperature of the working fluid is higher as possible. To increase the performance of heat exchangers and so reduce their size, Maizza and Maizza [3] suggest to use fluid with high density and high heat of vaporization. In 2012, Liu et al. [4], in the context of power generation, presented a two-stage Rankine cycle for electricity power plants. Ten pure working fluids from different chemical families (aromatics, hydrofluoroolefin, hydrofluorocarbons, hydrocarbons, ammonia) were tested. They concluded that the performance was not affected by the fluid selected. They concluded that selection would be realized on installation volume, size of the different components in the cycle, environmental protection, and operator safety. In 2017, Rahbar et al. [5] published a complete review of ORC for small-scale applications. In their paper, they present some main characteristic of the working fluid.

2. Description of Organic Rankine Cycle

An Organic Rankine Cycle is composed of a boiler/evaporator, a condenser, a pump, and a turbine (expander). The main application of ORC concerns the transformation/utilization of heat source at low or intermediate temperature (around 80°C). **Figure 1** reminds the schematic diagram of ORC. The pump compresses the liquid state working fluid until its desired pressure (and so temperature). The liquid is heated and vaporized in the boiler/evaporator which is also called the generator of vapor (the heat source). The vapor state working fluid is expanded in the turbine. During this operation electricity can be produced thanks to a generator. At low pressure, the working fluid is cooled in the condenser.

The cycle performance depends on the architecture of the system but also on the chosen working fluid and the operating conditions. The **Figure 2** presents a typical T-s diagram of an ORC with a heating and cooling medium.

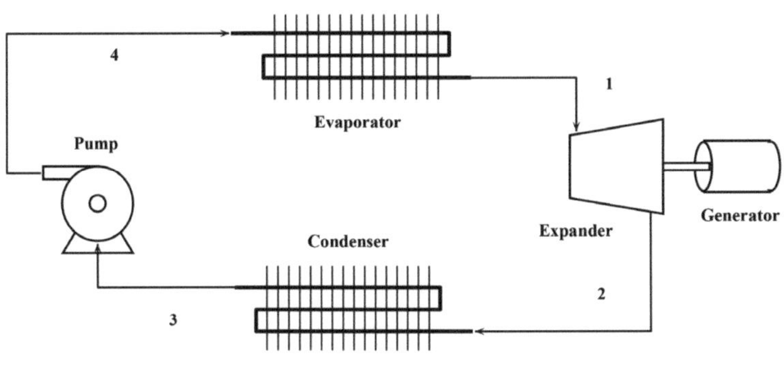

Figure 1.
Schematic diagram of ORC.

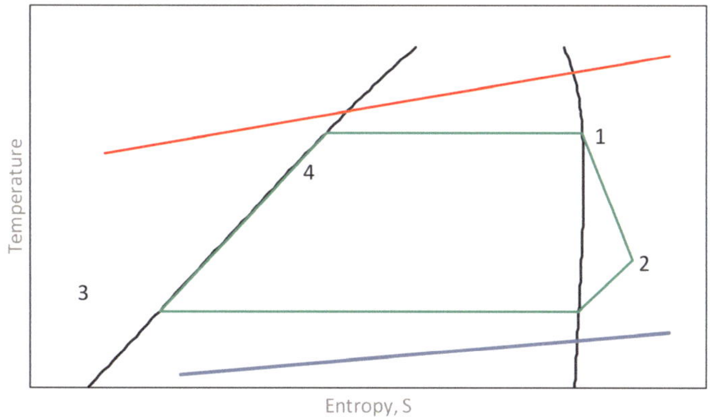

Figure 2.
Temperature-entropy diagram of the simple ORC. In green, working fluid; in red, heating medium; in blue, cooling medium.

3. Working fluid selection: characteristics

The choice of a working fluid is very crucial in the ORC systems. The working fluid should guarantee the cycle performance and system efficiency and must be adapted to the operating conditions of the system (temperature, pressure). Also hygiene, safety, and environment (HSE) aspects of the fluid have to be taken into account: the working fluid should satisfy the environment protection standards and ensure the safety of the operators. Moreover, the thermophysical properties of the working fluids impact strongly on component's size.

As the efficiency of the cycle is better for fluids with high critical temperature value [6], the working fluids are often heavy compounds with important molecular weight and low boiling point. Three types of organic fluids exist: dry, isentropic, and wet fluids. **Figure 3** presents the T-S diagrams of these fluids. For dry fluid, $dT/dS > 0$, and for isentropic fluid, $dT/dS = 0$. It signifies that during expansion, there is no formation of liquid droplet. For wet fluid, $dT/dS < 0$ (like water). During expansion, liquid droplets are formed and can damage the equipment. So, it can be necessary to add a superheating equipment.

In general, fluorinated components are used as working fluids. It is important to note that in 2014, European F-gas directive plans the prohibition of fluorinated working fluids with GWP of 2500 or more from 2020. The 2009 F-gas regulation fixes the limits of GWP for each year (Bolaji [7]). In 2018, the objective was to use fluids with GWP < 1300 close to the GWP of R134A (GWP = 1300). In 2020, the objective is to use fluids with GWP < 1000. In order to reach the objectives in terms of GWP, two solutions exist: the first one consists of the development of new fluids with low GWP values, such as hydrofluoroolefins (HFOs). The second one consists in developing new blends of refrigerants, less than four components [6]. With the existing equipment retrofit aspects are also very important. For all the cases, it is important to consider the thermophysical properties for the selection of the organic working fluid.

3.1 Chemical properties

Critical temperature: In order to guarantee ORC efficiency, the critical temperature of the chosen fluid should be as high as possible and close to the maximum temperature of the heat source.

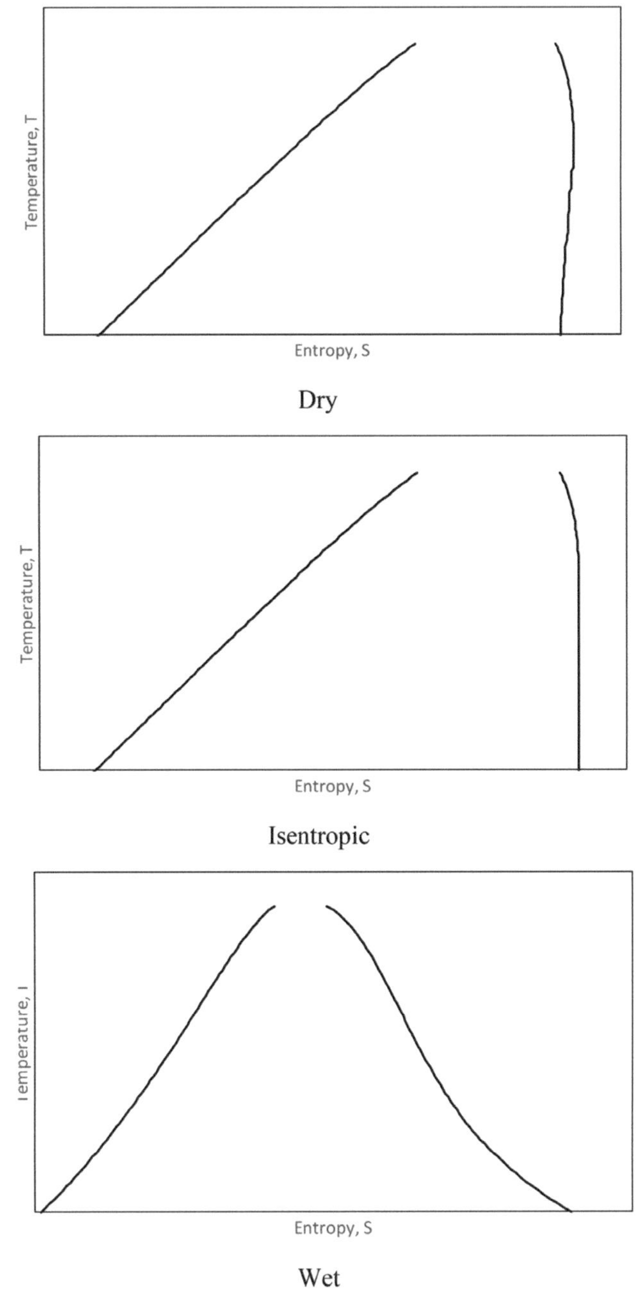

Figure 3.
Types of organic fluids (dry, isentropic, and wet) (Liu [6]).

Triple point temperature: As low as possible, the working fluid should not be freezing in ORC under low temperature condition (cold source).

Molecular weight: The larger is the molecular weight, the smaller is the specific enthalpy drop, and the lower is the number of stages required for the expander.

Thermal stability: The working fluid should not have the possibility to decompose itself under high pressure and high temperature conditions.

Safety and environment impacts: Nontoxic and non-flammable organic fluid is required to protect the personnel in case of fluid leakage. Also, the working fluid should have zero ozone depletion potential (ODP), minimal global warming potential (GWP), and low atmospheric lifetime (ALT).

Material compatibility: The working fluid should be non-corrosive and should have a non-eroding characteristic in order to guarantee the integrity of the components of the ORC.

3.2 Thermodynamic properties

Phase diagram: The knowledge of pure component phase diagram (mainly the pure component vapor pressure) is very important in order to know the physical state of the fluid. Concerning mixtures, the phase diagrams are more complex. Recently, Privat and Jaubert [8] have revisited the Scott and van Konynenburg [9] (**Figure 4**) classification of binary systems. Depending on the temperature and pressures conditions, vapor-liquid equilibrium but also vapor-liquid-liquid equilibrium (VLLE) and liquid-liquid equilibrium (LLE) can appear.

Density and specific volume: Working fluids with low specific volume lead to low volume flow rates and so have a non-negligible impact on the sizing of heat exchangers and expanders (low volume of vapor at the outlet).

Latent heat: In general, a working fluid with high heat of vaporization is preferable as more heat is absorbed during the evaporation step. But in case of the

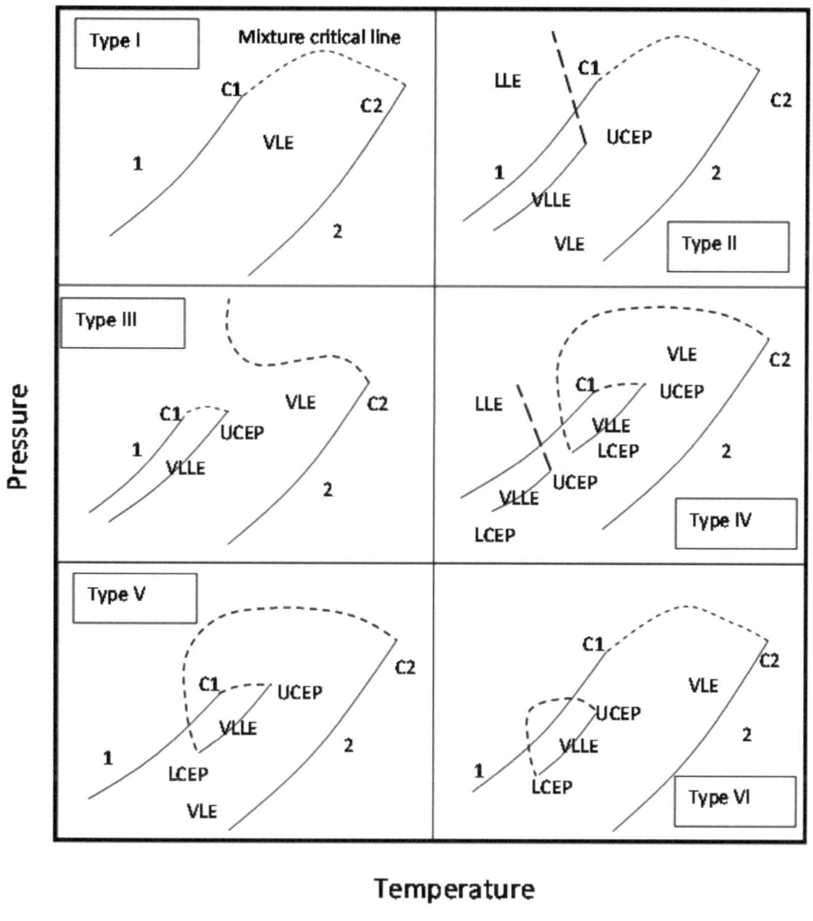

Figure 4.
Scott and van Konynenburg classification [9].

utilization of low-grade waste heat, it is better to use fluids with low latent heat of vaporization [5].

Speed of sound: The speed of the sound of the fluid limits the flow of fluid flowing in the expander. This parameter directly influences the size and therefore the cost of an ORC turbine.

The other thermodynamic properties are also useful but not crucial for the selection for the working fluid (heat capacity, surface tension).

3.3 Transport properties

Dynamic viscosity: Low dynamic viscosity for both liquid and vapor phases is preferable in order to reduce the friction losses and to maximize the heat transfer (reduction of the size of the heat exchangers, particularly heat surface exchange).

Thermal conductivity: High values of thermal conductivity are preferable in order to reduce the size of the heat exchangers (mainly surface of exchange).

4. Experimental techniques for the estimation of thermophysical properties

In this section, we will present several experimental techniques used for the experimental determination of thermodynamic and transport properties. Experimental methods for the investigation of thermophysical properties belong either to closed or open circuit methods.

Closed circuit methods: They can be divided into two main classes, depending on the method considered to determine the composition: static-analytic methods and static-synthetic methods. For the analytic methods, the composition of each phase is obtained by analyses after sampling (direct sampling method). For the synthetic methods, the global composition of the mixture is known a priori. No sampling is necessary.

Open circuit methods: There are several different techniques based on this principle. The mixture circulates through an equipment composed of sensitive element where the measurement of the thermophysical properties is realized. We can cite critical point measurement as an example of utilization of this technique. Also the densitometer technique developed by Galicia-Luna et al. [10] and Bouchot and Richon [11] which is a synthetic method can be classified as an open circuit method. A mixture with known composition circulates through a vibrating U-tube. Dynamic viscosity measurement can be also classified in this rubric.

4.1 Equilibrium properties and critical point

Synthetic or analytic methods can be considered for the determination of equilibrium properties. Vapor-liquid equilibrium properties can be obtained using the "static-analytic" method. Herein, the mixture is enclosed in an equilibrium cell equipped with a mixing mechanism to get fast equilibrium conditions. When the equilibrium is reached, small quantities of the phases are sampled and analyzed through chromatographic analyzers. A complete description of the setup is available in Wang et al.'s [12] paper. The apparatus is similar to the one present on **Figure 5**. Capillary samplers like ROLSI™ (Armines's patent) can be used to take samples.

The variable volume cell technique (**Figure 6**) can be cited as a static-synthetic method [13, 14]. The components of the mixture are introduced separately, and the composition is known by weighing procedure or after analysis. The volume of the cell is modified with a piston to study bubble points. At fixed temperature, saturating

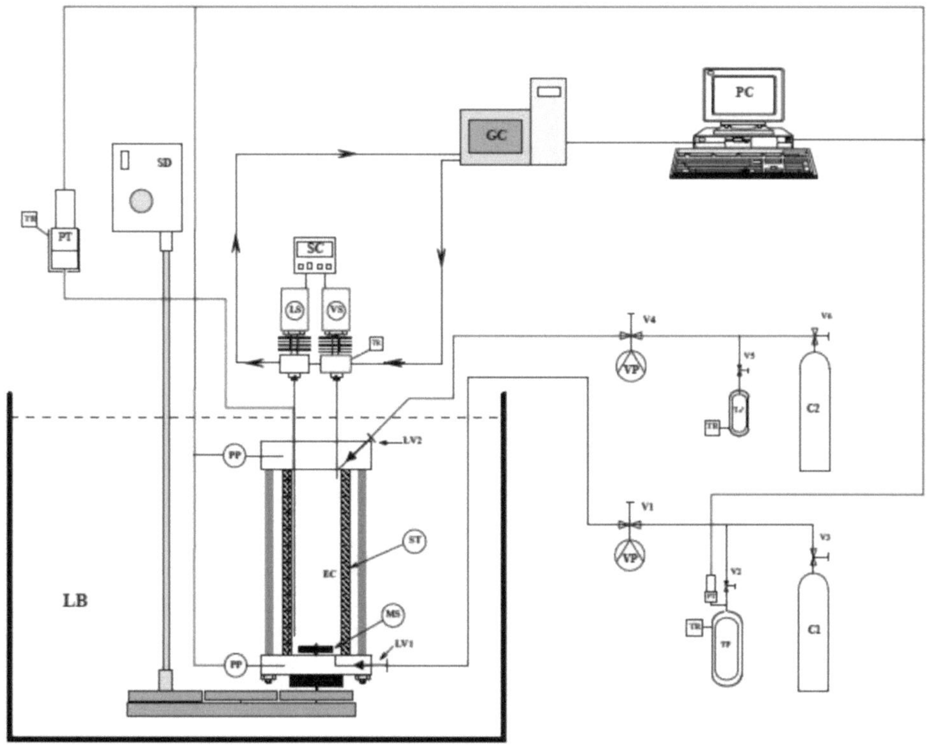

Figure 5.
Schematic diagram of the static-analytic apparatus. EC, equilibrium cell; LV, loading valve; MS, magnetic stirrer; PP, platinum resistance thermometer probe; PT, pressure transducer; RT, temperature regulator; LB, liquid bath; TP, thermal press; C1, more volatile compound; C2, less volatile compound; V, valve; GC, gas chromatograph; LS, liquid sampler; VS, vapor sampler; SC, sample controlling; PC, personal computer; VP, vacuum pump.

properties (pressure and saturated molar volume) of the mixture are determined through the pressure vs. volume curve recorded that displays a break point.

Isochoric method (**Figure 7**) can be used also to measure the dew point of multicomponent system. The components of the mixture are introduced separately, and the composition is known by weighing procedure or after analysis. The mixture is introduced at its vapor state, and then the temperature slowly decreases. The pressure and temperature are recorded. When the first drop of liquid appears, a break in the P-T curve is observed. The break point corresponds to the dew point. This technique is identical to the technique used to determine gas hydrate dissociation points [15].

Concerning critical point measurement, it is necessary to use a special device. The technique is based on dynamic-synthetic method where the mixture is circulating through the equilibrium cell (**Figure 8**) under specific conditions of temperature and pressure. A critical point can be determined by visually observing the critical opalescence and the simultaneous disappearance and reappearance of the meniscus, i.e., of the liquid-vapor interface from the middle of the view cell.

4.2 Volumetric properties or densities

It is well known that density is required for the development of equations of state and the development of models for mass and heat transfers. Several techniques which can be used to measure the density exist. We can cite the hydrostatic balance densitometer coupling with magnetic suspension (single-sinker method from Wagner et al. [16]), density measurement with vibrating bodies, bellows [17], and

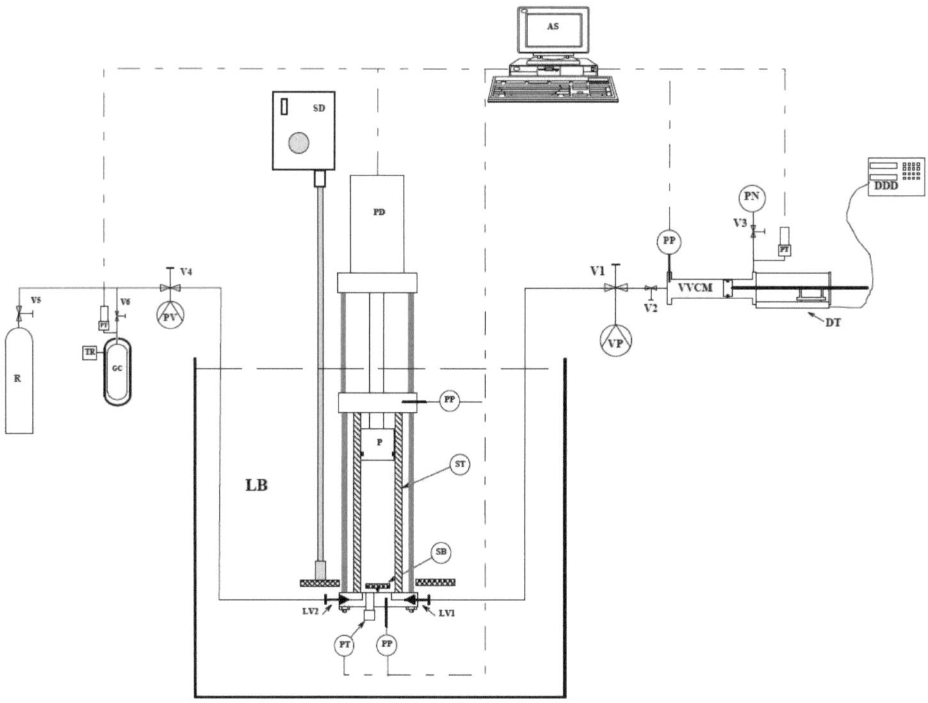

Figure 6.
Example of equipment which can be used for bubble point measurement. DAU, data acquisition unit; DDD, digital displacement display; DT, displacement transducer; GC, gas cylinder; LB, liquid bath; LVi, loading valve; P, piston; PM, piston monitoring; PN, pressurized nitrogen; PP, platinum probe; PT, pressure transducer; PV (VP), vacuum pump; R, gas reservoir; SD, stirring device; SB, stirring bar; ST, sapphire tube; TR, thermal regulator; Vi, valve; VVCM, variable volume cell.

isochoric method [18]. More details concerning these techniques of measurements (and others) are detailed in the IUPAC book dedicated to experimental thermodynamics [19]. Herein, we will only describe the technique based on vibrating tube densitometer.

A mixture with known composition can circulate through a vibrating U-tube (static or dynamic mode). Density is deduced from careful calibration using reference fluids. This apparatus can be used to obtain (PρT) data of compressed phases. A complete description of this technique is available in the papers of Coquelet et al. [20]. This technique is not very recommended close to the critical point. In effect, vibration of the tube may provoke a phase transition. Other techniques based on isochoric method can be used also to determine the volumetric properties at the vicinity of the critical point [21] (**Figure 9**).

4.3 Speed of sound

The speed of sound data is also very important to determine the equation of state, and it is linked to other thermodynamic properties. In effect, the isothermal compressibility κ_T is related to the isentropic compressibility κ_S via Maxwell's relations (Eq. (1)):

$$\kappa_T = \kappa_S + \alpha^2 v \frac{T}{C_p} \tag{1}$$

In Eq. (2), v [m^3/mol] is the molar volume of the compound, Cp is the heat capacity [J/mol], and κ_S is the isentropic compressibility $\kappa_S = -\frac{1}{v}\left(\frac{\partial v}{\partial P}\right)_S$ [Pa^{-1}] and is determined thanks to the measurements of speed of sound c [m/s] and density

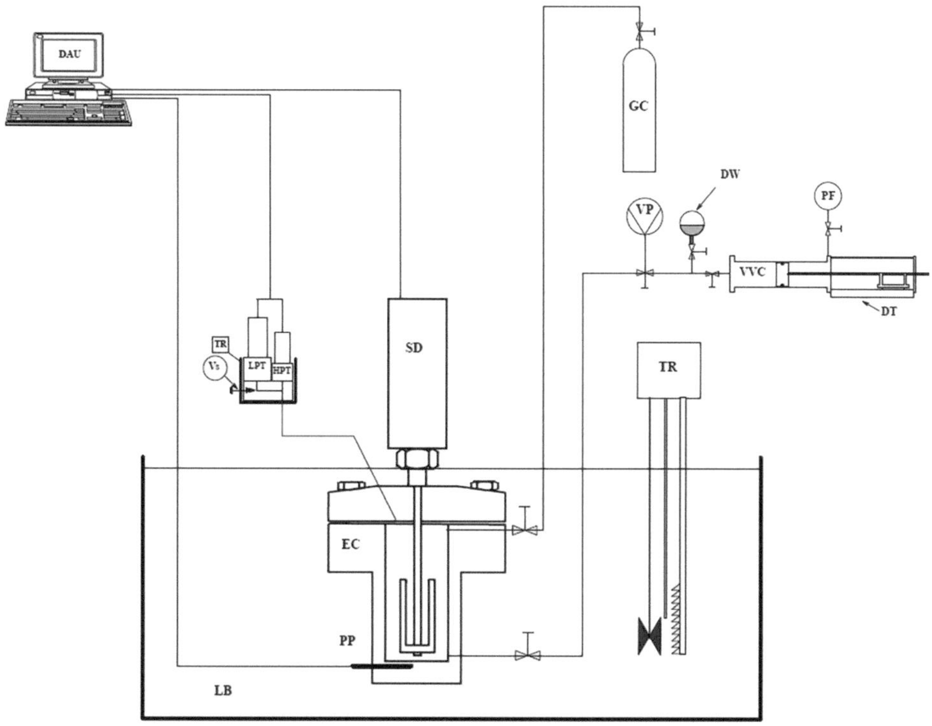

Figure 7.
Example of equipment which can be used for dew point measurement. DW, degassed water; DAU, data acquisition unit; EC, equilibrium cell; GC, gas cylinder; LPT, low pressure transducer; LPT, high pressure transducer; LB, liquid bath; PP, platinum probe; SD, stirring device; TR, temperature regulator; VP, vacuum pump; VVC, variable volume cell; PF, pressurizing fluid; DT, displacement transducer.

ρ [kg/m^3] reached in this work. Indeed, these three properties are linked in the liquid phase via $c = \sqrt{\frac{1}{\kappa_S \rho}}$. Like with density measurement, several techniques which were developed to obtain the value of speed of sound exist. The spherical resonator developed by Mehl and Moldover [22] (with high accuracy in gases), Trusler and Zarari [23], and Benedetto et al. [24] can be cited as example. For liquid and dense fluid, pulse-echo techniques are preferred for measuring the speed of sound particularly at high pressure. More details concerning these techniques of measurements (and others) are detailed in the IUPAC book dedicated to experimental thermodynamics [19]. Herein, we will describe one technique used at Heriot-Watt University [25]. **Figure 10** describes the cell of measurement. A cylindrical acoustic cell with well-known dimension is considered to measure the sound speed in the fluid using through-transmission method of ultrasonic testing. In this method, a transducer is located on one side of the cell, and a detector is placed on the opposite side of the acoustic cell (electric signal is converted into ultrasound waves and vice versa). An oscilloscope is used to observe the waves. Speed of sound is obtained by dividing the period of the waves by the distance between the speed of sound transducer and detector.

4.4 Heat capacity

The determination of isobaric heat capacity is done using a differential scanning calorimeter (DSC). The equipment is composed of two cells: one is the measurement cell, and one is the reference cell. A sample is introduced into the measurement cell, and a temperature ramp is applied. Knowing the heat flux transferred (absorbed or released, $\phi = \left(\frac{\partial H}{\partial t}\right)$) and the ramp, it is possible to estimate the heat capacity (Eq. (2)):

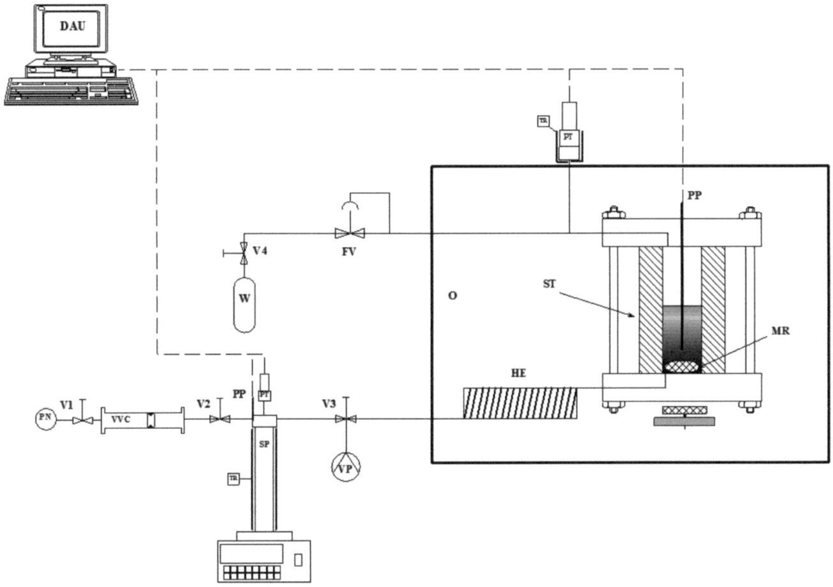

Figure 8.
Schematic diagram of the critical point measurement apparatus. DAU, data acquisition unit; FV, flow regulation valve; HE, heat exchanger; MR, magnetic rode, O, oven; PN, pressurized nitrogen; PP, platinum probe; PT, pressure transducer; ST, sapphire tube; SP, syringe pump; TR, temperature regulator; Vi, valve; VVC, variable volume cell; VP, vacuum pump; W, waste.

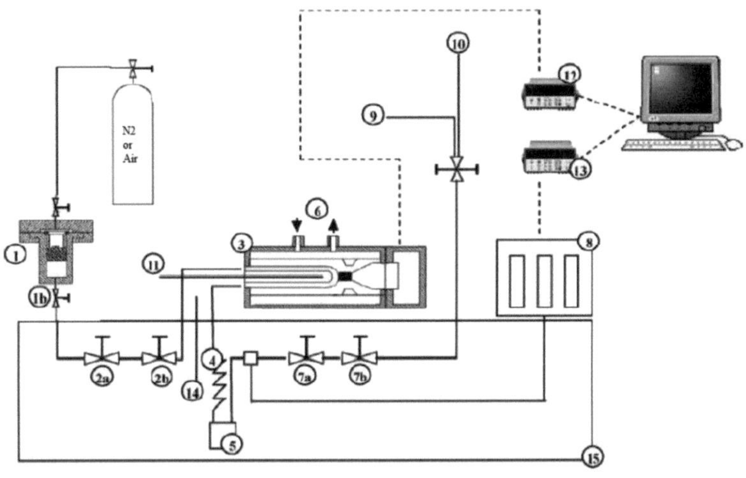

Figure 9.
Flow diagram of the vibrating tube densimeter [20]. (1) loading cell, (2a and 2b) regulating and shutoff valves, (3) DMA-512P densimeter, (4) heat exchanger, (5) bursting disk, (6) inlet of the temperature regulating fluid, (7a and 7b) regulating and shutoff valves, (8) pressure transducers, (9) vacuum pump, (10) vent, (11) vibrating cell temperature probe, (12) HP 53131A data acquisition unit, (13) HP34970A data acquisition unit, (14) bath temperature probe, (15) principal liquid bath.

$$C_P = \phi \Big/ \left(\frac{\partial T}{\partial t}\right) \tag{2}$$

In Eq. (4), H is the enthalpy, T is the temperature, and t is the time. A complete description of this technique is available in the paper of Al Ghafri et al. [26]. **Figure 11** presents a short description of the technique.

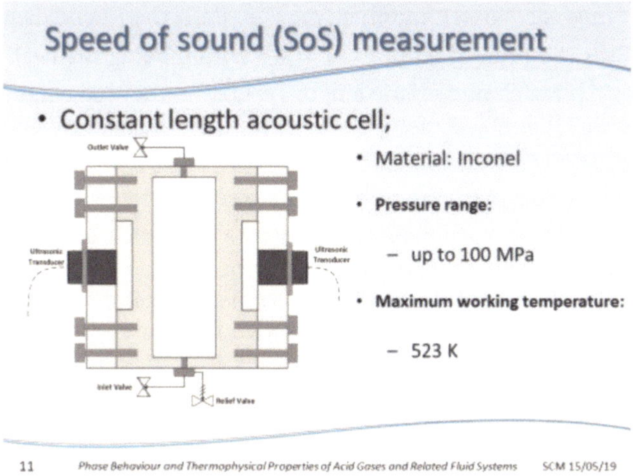

Figure 10.
View of the acoustic cell of equipment designed by Ahmadi et al. [25].

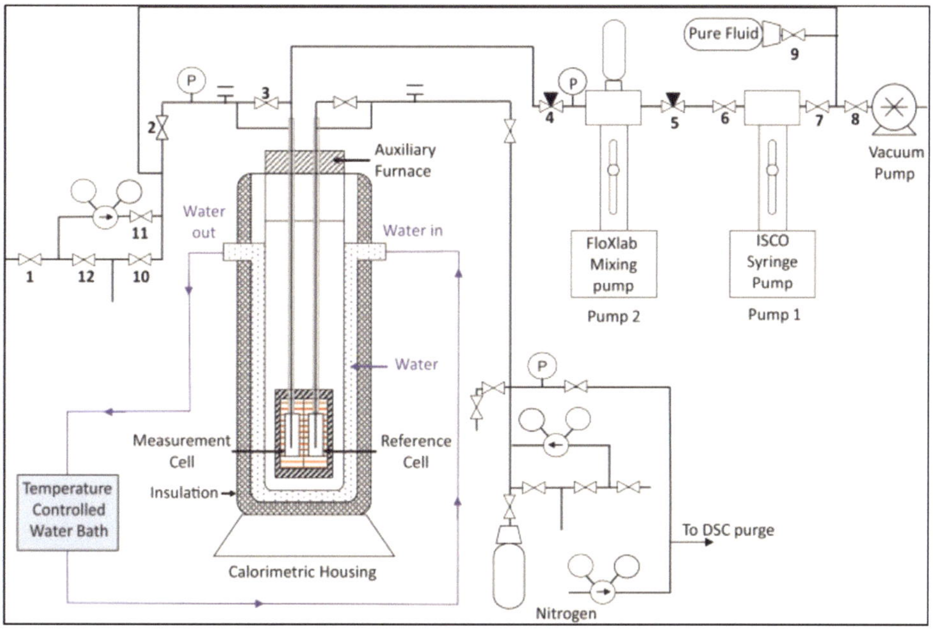

Figure 11.
Differential scanning calorimeter used for measurements of the isobaric heat capacities of liquid refrigerant mixtures from Al Ghafri et al. [26].

4.5 Transport properties

4.5.1 Interest of transport properties in engineering

Chemical and mechanical engineers use correlations with nondimensional numbers (Eq. (3)) to estimate heat transfer coefficient (h) essential for the estimation of global heat transfer coefficient of the heat exchanger and so the design of heat exchangers:

$$Nu = A\mathrm{Re}^{\alpha}\mathrm{Pr}^{\beta} \tag{3}$$

In Eq. (1), Nu is the Nusselt number $Nu = \frac{hL}{\lambda}$, Re is the Reynolds number $Re = \frac{\rho v L}{\mu}$, and Pr is the Prandtl number $Pr = \frac{C_p \mu}{\lambda}$ with h being the heat transfer coefficient [J.m^{-2}], L a characteristic length [m], λ the thermal conductivity [J. m^{-1}], ρ the density [kg.m^3], μ the viscosity [Pa.s], Cp the heat capacity [J.mol^{-1}. K^{-1}], and v the speed of the fluid [m.s^{-1}]. The correlation can be modified with the utilization of Weber number ($We = \frac{\rho L v^2}{\gamma}$, γ is the surface tension) taking into account the effect of interfacial tension during the formation of the bubble of gas in the evaporator or drop of liquids in the condenser.

We remind that for a heat exchanger, global heat transfer coefficient (U) depends on the local heat transfers (h) of the two fluids and the thermal conductivity of the material of the heat exchanger.

4.5.2 Dynamic viscosity

Like density measurements, several techniques to determine viscosity of fluids exist. For example, we can cite the falling ball technique [27], capillary technique, vibration quartz [28], and vibrating bodies [29, 30]. We invite the reader to have a look in the paper from Le Neindre [31] for a complete overview of the techniques of measurement. Herein we will only describe the viscometer used to measure dynamic viscosities using the capillary tube viscosity method. A schematic view of the setup used at Heriot-Watt University is shown in **Figure 12**. The apparatus is

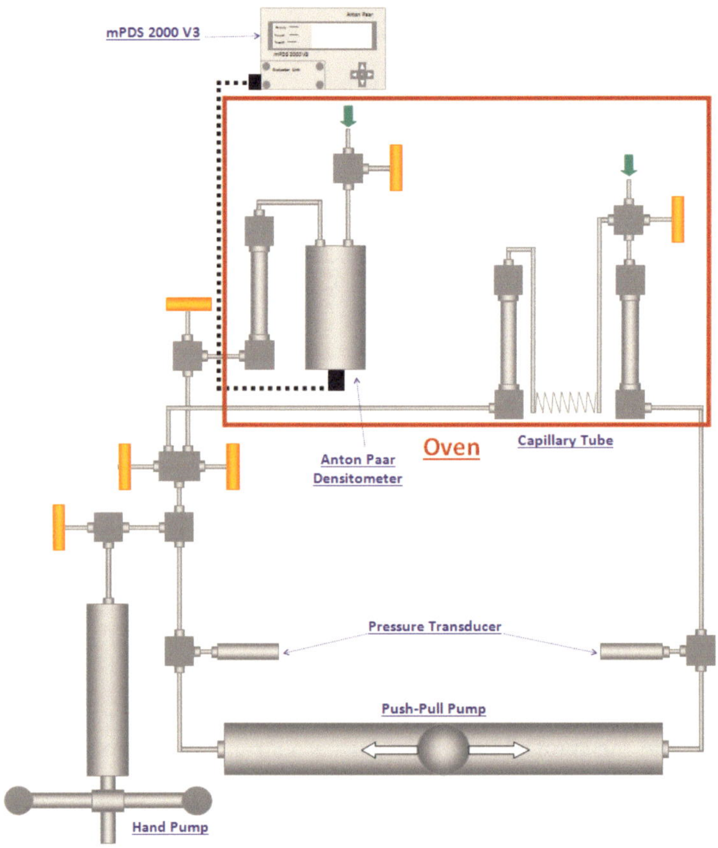

Figure 12.
Schematic diagram of the setup used for dynamic viscosity measurement [32].

comprised of two small cylinders connected to each other through a capillary tube and a temperature-dependent calibrated internal diameter. A complete description of this technique is available in the paper of Kashefi et al. [32]. The temperature of the system was set to the desired condition, and the desired pressure was set using the hand pump. Poiseuille equation (Eq. (4)) (in laminar flow conditions) can relate the pressure drop across the capillary tube to the viscosity, tube characteristics, and also flow rate for laminar flow:

$$\Delta P = \frac{128LQ\mu}{C\pi D^4} \tag{4}$$

In Eq. (3), ΔP is the differential pressure across the capillary tube viscometer in psi, Q represents the flow rate in cm^3/sec, L is the length of the capillary tube in cm, D refers to the internal diameter of the capillary tube in cm, μ is the dynamic viscosity of the flown fluid in cP, and C is the unit conversion factor equal to 6,894,757 if the above units are used.

4.5.3 Thermal conductivity

Several techniques exist to measure the thermal conductivity of fluids [33]. Among those methods, the transient hot-wire method is considered to be the most used technique and to be a very accurate and reliable technique to measure this thermophysical property. The basic theory of the transient hot-wire method is presented in Healy's paper [34], and procedure of measurement is fully described in the paper of Marsh et al. [35]. Generally, a transient hot-wire apparatus consists of one highly pure platinum wire, a current source, a voltage meter, a data acquisition system, and a computer (**Figure 13**). The current source provides a constant current for the platinum wire, which is embedded in the tested fluid. Then the temperature of the wire will rise because of the Joule effect. Subsequently, the temperature of the

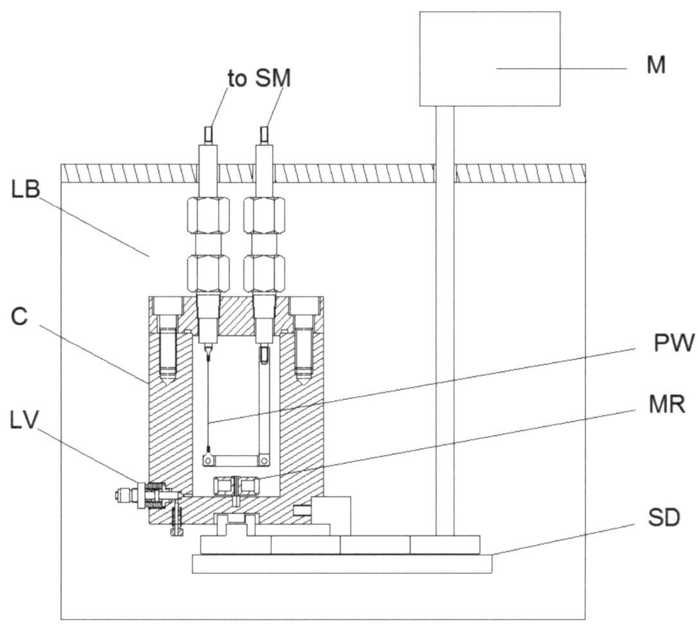

Figure 13.
View of the thermal conductivity cell (hot-wire method). LB, liquid bath; C, cell; LV, loading valve; M, motor; PW, platinum wire; MR, magnetic rod; SD, stirring device; to SM, to SourceMeter.

fluid will also change as a result of the heat conduction between the wire and the fluid (heat radiation and convection are in general neglected). The temperature rise of the platinum wire as a function of the duration t is given by Eq. (5):

$$\Delta T = \frac{q}{4\pi\lambda} \ln\left(\frac{4\kappa t}{r^2 C}\right) \tag{5}$$

In Eq. (5), λ is the thermal conductivity of the sample, q is the constant power provided to the wire, W/m. κ is the thermal diffusivity of the fluid, r is the radius of the hot wire, and C is Euler's constant, whose value is 1.781. As the relation between the ΔT and ln(t) can be determined through the experiments, the thermal conductivity of the tested fluid can be calculated using this equation. Compared with other methods, the convenience, accuracy, and the short duration of transient hot-wire method make it a widely used method nowadays.

5. Data treatment

Process or system simulator required models or correlation in order to define the best operating conditions (T, P, flow, compositions) with the maximum of efficiency and the best coefficient of performance (COP) but also to design each component (heat exchangers, expanders, valves, expander). In this section we will just present few models which can be used with the experimental data in order to predict the phase diagrams and thermodynamic and transport properties.

5.1 Thermodynamic models

In this section, we propose a presentation of three types of thermodynamic models. These models are briefly described. The main difference concerns the number of parameters we have to adjust and so the quantity and type of experimental data we have to acquire in order to adjust the parameters.

The thermodynamic properties are obtained by using equations of state (EoSs). In the process simulators, the most famous EoSs are of cubic type, such as Eq. (6):

$$P = \frac{RT}{v-b} - \frac{a\alpha(T)}{v^2 + uvb + wb^2} \tag{6}$$

In Eq. (6), R is the ideal gas constant, a and b are the parameters of the EoS calculated using the critical temperature (Tc) and pressure (Pc) of each component, and α(T) is a function of temperature, acentric factor, Tc, and Pc. u and w are other parameters.

Another type of equation of state of the molecular type can be used. The Helmholtz energy is calculated by considering all the molecular interactions like dispersion, polarity, H bonding (association), etc. Equation (7) describes the method of calculation of the Helmholtz energy A:

$$\frac{A}{Nk_bT} = \frac{A^{\text{Ideal}}}{Nk_bT} + \frac{A^{\text{Segment}}}{Nk_bT} + \frac{A^{\text{Chain}}}{Nk_bT} + \frac{A^{\text{Association}}}{Nk_bT} \tag{7}$$

In Eq. (7), k_b is the Boltzmann constant, T is the temperature, and N is the mole number. The most known molecular EoSs of this type are SAFT type. Based on the Wertheim's statistical theory of associative fluids (1984), Chapman et al. [36]

developed the first EoS SAFT (statistical associating fluid theory) called SAFT-0. Many versions exist today, such as SAFT-VR [37], PC-SAFT [38], and SAFT Mie [39]. The various versions differ mainly in the choice of the reference fluid, the radial distribution function, and explicit expressions of the terms of perturbation.

The last type of equation of state which can be used to estimate the thermodynamic properties is based on multi-fluid approximation. It is well known that from the knowledge of Helmholtz energy, all thermodynamic properties can be calculated. This equation of state is explained in terms of reduced molar Helmholtz free energy (Eq. (8)). Temperature and density are expressed in the dimensionless variables $\delta = \frac{\rho}{\rho_C}$ and $\tau = \frac{T}{T_C}$:

$$\frac{A(T_r, \rho_r, \overline{x})}{RT} = a(T_r, \rho_r, \overline{x}) = a^{id}(T_r, \rho_r, \overline{x}) + a^{res}(T_r, \rho_r, \overline{x}) \tag{8}$$

In Eq. (8), the exponent "id" stands for the ideal gas contribution, and exponent "res" is the residual contribution. ρ_r and T_r are the reduced density and temperature, respectively. Concerning the development of equation of state for the mixtures, the first possibility is to consider mixing rules for each parameter like in the cubic equation of state. The best approach is to consider the multi-fluid approximation. This approach was introduced by Tillner-Roth in 1993 [40]. It applies mixing rules to the Helmholtz free energy of the mixture of components (Eq. (9)):

$$\begin{aligned} a(T_r, \rho_r, \overline{x}) = \frac{A(T_r, \rho_r, \overline{x})}{RT} &= \sum_j x_j \left(a_j^{id}(T_r, \rho_r) + a_j^{res}(T_r, \rho_r) \right) + x_j \ln x_j \\ &+ \sum_{p=1} \sum_{q=p+1} x_p x_q F_{pq} a_{pq}^E \end{aligned} \tag{9}$$

In Eq. (9), superscript "E" is for excess properties, and subscripts "p," "q," and "j" are the component index. An excess property is defined to calculate the deviation from ideal mixture. $\Delta a_{pq}^{res} = \sum_{p=1} \sum_{q=p+1} x_p x_q F_{pq} a_{pq}^E$ is called departure function from the ideal solution. It is an empirical function, like Eq. (10), fitted to experimental binary mixture data, mainly densities, speed of sound, or heat of mixing. In this departure function, the F_{pq} parameters take into account the behavior of one binary pair with another. If only vapor-liquid equilibrium properties are available, a_{pq}^E can be considered to be equal to zero [41]:

$$\frac{A(T_r, \rho_r)}{RT} = \ln(\delta) + \sum_i \alpha_i \tau^{t_i} + \sum_k \alpha_k \tau^{t_k} \delta^{d_k} \exp\left(-\gamma \delta^{l_k}\right) \tag{10}$$

Equation (10) contains several adjustable parameters α_i, α_k, t_k, d_k, and l_k. Experimental data is required to adjust these parameters. Moreover, if $l_k = 0$, then $\gamma = 0$ and if $l_k \neq 0$, then $\gamma = 1$.

With the multi-fluid approximation, it is important to calculate the new critical properties corresponding to the mixture studied (superscript "mix") because reduced parameters are used in Eq. (8).

For example, we can use $T_C^{mix} = \sum_{p=1} \sum_{p=1} k_{T,pq} x_p x_q \left(T_{Cp} T_{Cq}\right)^{0.5}$ and $v_C^{mix} = \sum_{p=1} \sum_{p=1} k_{v,pq} x_p x_q \frac{1}{8} \left(\left(v_{Cp}\right)^{1/3} + \left(v_{Cq}\right)^{1/3} \right)^3$ where $k_{T,pq}$ and $k_{v,pq}$ are adjustable binary interaction parameters. VLE data should be used to fit $k_{T,pq}$ parameters, and if experimental data concerning densities of mixture are available, it is possible to fit $k_{v,pq}$.

5.2 Transport property models

Concerning the transport properties, different approaches exist. One consists in using the corresponding state method. The most famous approach is the TRAPP method developed by the NIST. Huber et al. [42] and Klein et al. [43] have developed a series of equations adapted to the prediction of viscosities and thermal conductivities of pure components and mixtures. The approach consists in modifying the transport properties in the ideal dilute gas state taking into account the molecular interactions (and so the density of the fluid with temperature and pressure).

The viscosity of a dilute gas can be determined as a function of temperature by Eq. (11). A dilute gas is composed by noninteractive rigid spheres of diameter σ:

$$\eta^{\circ} = C\frac{T^{\frac{1}{2}}M^{\frac{1}{2}}}{\sigma^{2}} \tag{11}$$

In Eq. (11), T is temperature in K, M is the molar mass in g/mol, σ is the diameter of the rigid sphere in nm, and C is a constant depending of the molecule considered. Chapman-Enskog cited in Poling and Prausnitz [44] have introduced an additional term called "collision integral" which takes into account the collision between the molecules (Eq. (12)):

$$\eta^{\circ} = K\frac{T^{\frac{1}{2}}M^{\frac{1}{2}}}{\sigma^{2}\Omega} \tag{12}$$

In Eq. (12), $\ln(\Omega) = \sum_i a_i\left(\ln\left(\frac{T}{\frac{\epsilon}{k}}\right)\right)^i$, and ?/k is an empirical factor link to the potential of interaction.

Concerning thermal conductivity, $\lambda_0(T)$ represents the dilute gas thermal conductivity and can be calculated by Eq. (13):

$$\lambda_0 = A_1 + A_2(T/T_C) + A_3(T/T_C)^2 \tag{13}$$

When the pressure and so the density of the fluid increases, molecular interaction between molecules cannot be neglected, and so it is necessary to apply a correction (like a residual term for equation of state). Equations (14) and (15) lead to calculate dynamic viscosity and thermal conductivity:

$$\eta = \eta^{\circ} + \Delta\eta \tag{14}$$

In Eq. (14), $\Delta\eta$ is the residual viscosity:

$$\lambda(\rho, T) = \lambda_0(T) + \Delta\lambda_r(\rho, T) + \Delta\lambda_C(\rho, T) \tag{15}$$

In Eq. (15), $\Delta\lambda_r(\rho, T)$ is the residual thermal conductivity, and $\Delta\lambda_C(\rho, T)$ is the empirical critical enhancement.

The R134a is the reference fluid for the refrigerants [42], but for the hydrocarbons, it is better to consider propane or methane [45]. The viscosity and thermal conductivity of the other fluids are calculated from the properties of reference fluids. Critical properties of the fluid are required. Equation (16) presents the equation for the dynamic viscosity:

$$\eta(T) = \eta^{\circ}(T) + \Delta\eta^R\left(\frac{T}{f}, \rho h\right)F_\eta \tag{16}$$

In Eq. (16), $f = \frac{T_c}{T_c^R}$, $h = \frac{\rho_c^R}{\rho_c}$, and $F_\eta = f^{\frac{1}{2}}h^{-\frac{2}{3}}\left(\frac{M}{M^R}\right)^{\frac{1}{2}}$ reduction ratio. Equation (17) presents the equation for the thermal conductivity:

$$\lambda(T) = \lambda^{\circ}(T) + \Delta\lambda^{R}\left(\frac{T}{f}, \rho h\right)F_{\lambda} \tag{17}$$

In Eq. (17), $f = \frac{T_c}{T_c^R}$, $h = \frac{\rho_c^R}{\rho_c}$, and $F_\lambda = f^{\frac{1}{2}}h^{-\frac{2}{3}}\left(\frac{M^R}{M}\right)^{\frac{1}{2}}$ reduction ratio. For mixtures, mixing rules have to be considered [46].

6. Presentation of some results

In this section, we will present some results obtained for pure component (transport properties) and multicomponent systems (equilibrium properties) on working fluids already published.

6.1 Phase diagram

We will present the results obtained for three binary systems. The first one concerns a mixture of CO_2 and R1234ze(e) published by Wang et al. [12]). **Figure 14** presents the phase diagram. The system presents no azeotropic behavior. We can notice that with the experimental data, we can also predict a critical point using asymptotic laws of behavior [12]. **Figure 15** presents a comparison with the critical point measured by Juntaratchat et al. [50] using a critical point setup similar to the equipment already presented in Section 4.

The binary system R245fa + isopentane is presented on **Figure 16** at 392.87 K. Measurement was done using static-analytic method [51]. We have used REFPROP 10.0 to correlate the data (REFPROP 10.0 [52] uses Fundamental Helmholtz equation). We can observe a good agreement between REFPROP prediction and the experimental data (we have one comment concerning this point: the reference of the data used in the data treatment by REFPROP is not clearly mentioned).

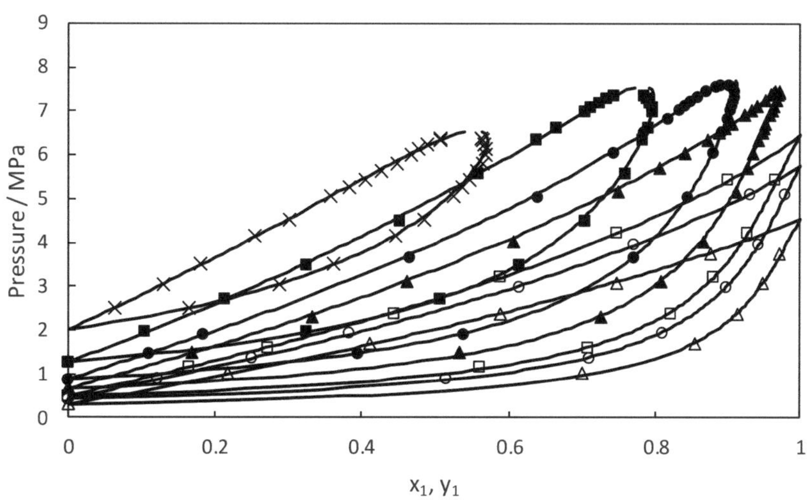

Figure 14.
Pressure as a function of CO_2 mole fraction in the CO_2. (1)−R-1234ze(E) and (2) mixture at different temperatures. Δ, 283.32 K; o, 293.15 K; □, 298.15 K; ▲, 308.13 K; ●, 318.11 K; ■, 333.01 K; ×, 353.02 K. Solid lines: calculated with Peng-Robinson EoS [47], Wong-Sandler mixing rules [48], and NRTL [49] activity coefficient model [12].

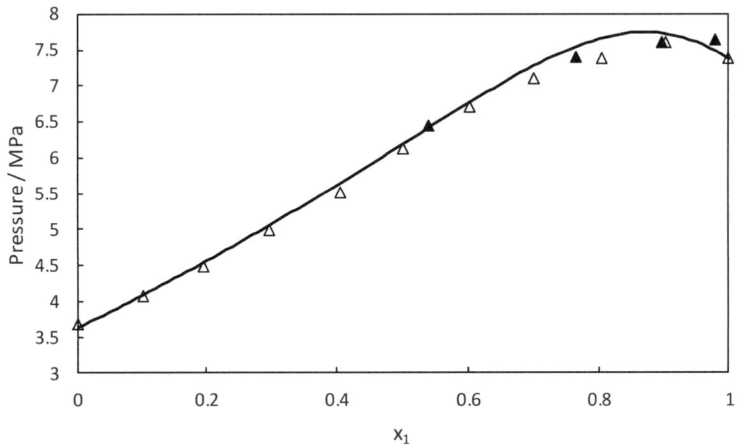

Figure 15.
Critical pressure of the binary system CO2 (1) +R-1234ze(E) (2) [12]. Solid line: calculated using PR EoS, Wong-Sandler mixing rules, and NRTL activity coefficient model. Δ, experimental data from Juntaratchat et al. [50]; ▲, predicted using power laws with asymptotic behavior at critical point.

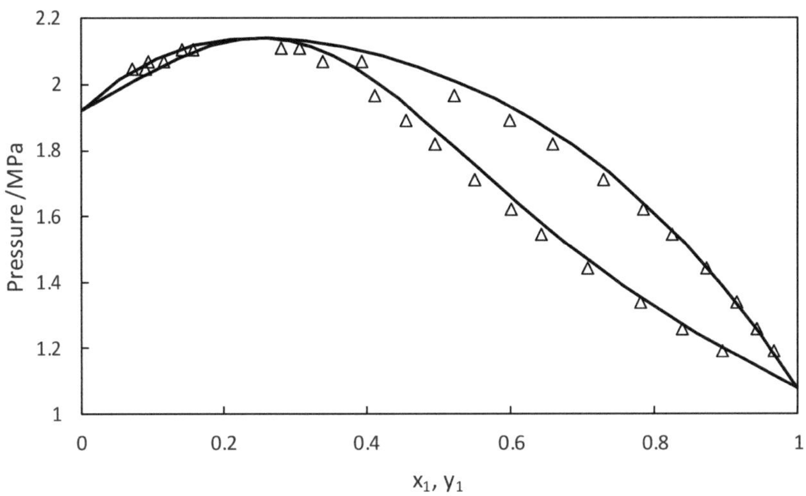

Figure 16.
Pressure as a function of CO2 mole fraction in the isopentane. (1) –R245fa and (2) mixture at 392.87 K. Comparison between experimental data [51] and modeling using REFPROP 10.0.

6.2 Bubble point

Variable volume cell was used to determine bubble pressure of the ternary system composed with R32 + R290 + R227ea [53]. **Figure 17** presents the results obtained. The data were correlated by a cubic equation of state (Redlich-Kwong-Soave EoS [54], MHV1 [55] mixing rules, and NRTL [49] activity coefficient model).

6.3 Densities

Vibrating tube densitometer technique was used to measure the densities of the R1216 [20]. The results are presented on **Figure 18**. The same technique is used to obtain density data of R1234yf (**Figure 19**) at saturation [56]. Density at the vicinity

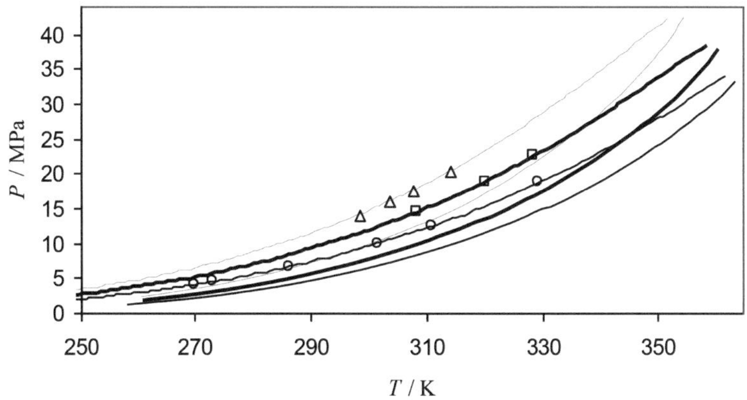

Figure 17.
The system R32 (1) +R290 (2) +R227ea (3) [53]. Pressure versus temperature diagram for each composition. Mixture 1: ×1 = 0.322, ×2 = 0.123, (□) experimental bubble points. Mixture 2: ×1 = 0.135, ×2 = 0.174, (○) experimental bubble points. Mixture 3: ×1 = 0.493, ×2 = 0.127, (Δ) experimental bubble points.

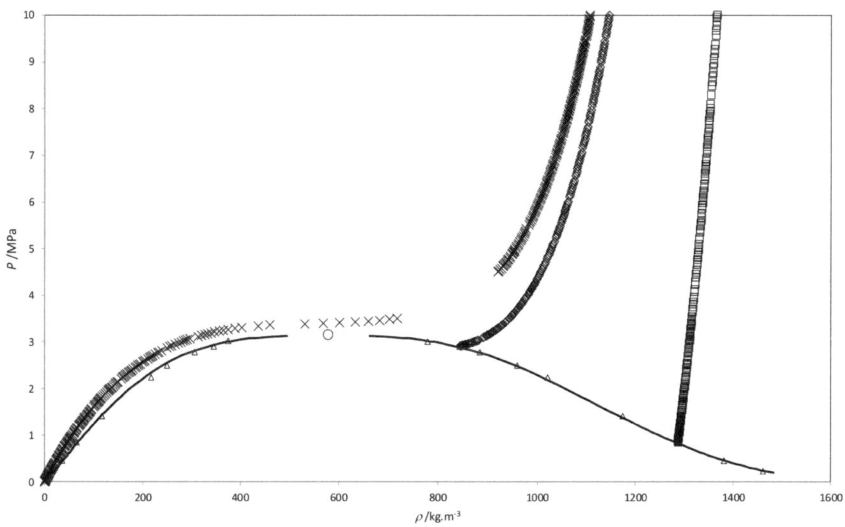

Figure 18.
Pressure density phase diagram of hexafluoropropene (R1216). (Δ), experimental densities at saturation; out of saturation; (×) 362.90 K; (◇) 355.18 K and (▯) 303.28 K; O, critical point [20].

of the critical point was obtained by an isochoric method used by Tanaka and Higashi [21].

6.4 Dynamic viscosity

Capillary viscometer was used by Laesecke et al. [57] to determine the dynamic viscosity of liquid R245fa at saturation. **Figure 20** presents the results and comparison with prediction using REFPROP 10.0. A good agreement is observed.

6.5 Thermal conductivity

Marrucho et al. [58] have used the transient hot-wire technique to measure the thermal conductivity of R365mfc. **Figure 21** presents the results obtained at 336.85

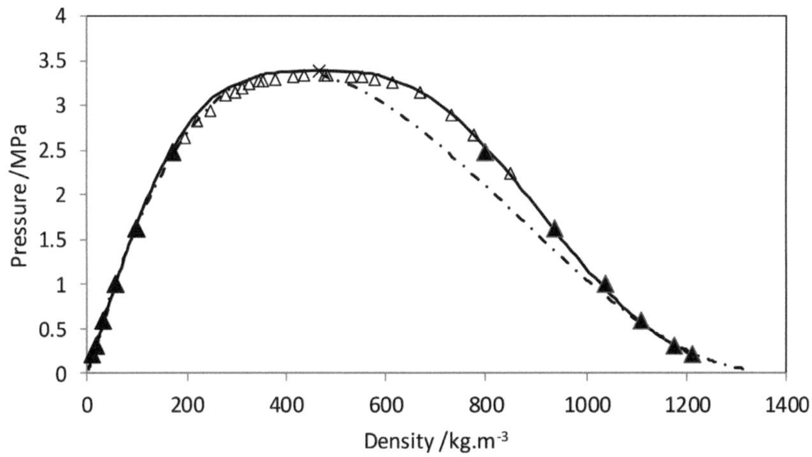

Figure 19.
Pressure density phase diagram of R1234yf at saturation. (Δ) data from Tanaka and Higashi [X], (▲) Coquelet et al. [56], (×) critical point from REFPROP. Solid line, correlation presented in Coquelet et al. [20]; (—·), Peng-Robinson EoS.

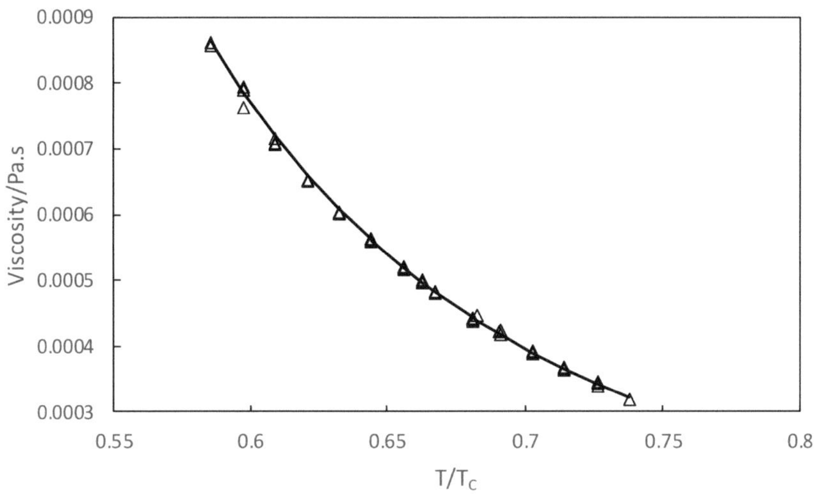

Figure 20.
Dynamic viscosity of R245fa at saturation. Symbol: experimental data from Laesecke et al. [57]. Solid line: prediction using REFPROP 10.0.

and 377.4 K for several pressures from 1 to 4.5 bar. Comparison was done with prediction from REFPROP 10.0. A good agreement is observed.

7. Conclusion

The optimization of ORC depends strongly on the capability of models to predict the thermophysical properties of working fluids for their selection and retrofit in existing ORC equipment. The main thermophysical properties include phase equilibria (and so critical point), densities, speed of sound, dynamic viscosity, and thermal conductivity. Through this book chapter, the reader can easily understand that several experimental techniques developed to measure the thermophysical properties exist. Some of them were presented and described. In general, these

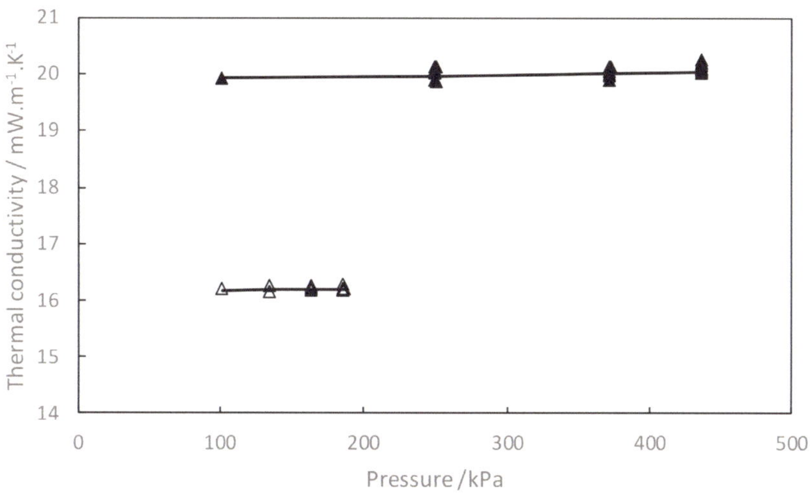

Figure 21.
Thermal conductivity of R365mfc as a function of pressure. Symbol: experimental data from Marrucho et al. [58] at (Δ) 336.85 K and (▲) 377.40 K. Solid line: prediction using REFPROP 10.0.

techniques are very accurate, and we recommend to use the experimental technique with which the operator/engineer is the most familiar with. It is obvious to see that we have the capability to determine all the thermophysical properties of interest for the design and optimization of ORC. In our opinion, there are two main challenges for the future. The first one concerns the development of equipment which permits the determination of several thermophysical properties together and particularly thermophysical properties at equilibrium. The second one concerns the development of equipment which require a small quantity of sample and so the utilization of in situ analysis techniques. In effect, it is difficult to have important quantity of new synthesis molecules (with high purity) to realize the measurement of accurate thermophysical properties of pure components and mixtures. The experimental data will be used to adjust parameters on models or correlations. Data acquisition is an essential step for a good working fluid design. In the literature an optimal selection (**Figure 22**) of working fluids as a function of the temperature of the heating medium exists [59]. This classification does not concern mixtures, and in the case of the utilization of mixtures, experimental characterization has to be done.

Temperature /K						
320	365	395	420	445	465	500
R134a	R22	R152a	R600a	R600	R123	
R32	R290	R124	R142b	R254fa	R365mfc	
	R134a	CF3I	R236a	Néopentene	R601a	
		R236fa	Isobutene	R245ca	R601	
			Butene		R141b	

Figure 22.
Some working fluid for several level of temperature of heating medium [59].

Author details

Christophe Coquelet*, Alain Valtz and Pascal Théveneau
CTP-Centre of Thermodynamics of Processes, Mines ParisTech, PSL University, France

*Address all correspondence to: christophe.coquelet@mines-paristech.fr

References

[1] Badr O, Probert SD, O'Callaghan PW. Selecting a working fluid for a Rankine cycle engine. Applied Energy. 1985;**21**:1-42. DOI: 10.1016/0306-2619(85)90072-8

[2] Saleh B, Kogalbauer G, Wendland M, Fischer J. Working fluids for low temperature organic Rankine cycles. Energy. 2007;**32**:1210-1221. DOI: 10.1016/j.energy.2006.07.001

[3] Maizza V, Maizza A. Unconventional working fluids in organic Rankine Cycles for waste energy recovery systems. Applied Thermal Engineering. 2001;**21**:381-390. DOI: 10.1016/S1359-4311(00)00044-2

[4] Liu B, Riviere P, Coquelet C, Gicquel R, David F. Investigation of a two stage Rankine cycle of electricity power plants. Applied Energy. 2012;**100**: 285-294. DOI: 10.1016/j.apenergy.2012.05.044

[5] Rahbar K, Mahmoud S, Al-Dadah RK, Moazami N, Mirhadizadeh SA. Review of organic Rankine cycle for small-scale applications. Energy Conversion and Management. 2017;**134**: 135-155. DOI: 10.1016/j.enconman.2016.12.023

[6] Liu B. Modélisation d'un cycle de production d'électricité bi-étagé à aeroréfrigérant sec [thesis]. MinesParisTech; 2014

[7] Bolaji BO. Experimental study of R152a and R32 to replace R134a in a domestic refrigerator. Energy. 2010; **35**(9):3793-3798. DOI: 10.1016/j.energy.2010.05.031

[8] Privat R, Jaubert JN. Classification of global fluid-phase equilibrium behaviors in binary systems. Chemical Engineering Research and Design. 2013; **91**:1807-1839. DOI: 10.1016/j.cherd.2013.06.026

[9] Scott RL, van Konynenburg PH. Critical lines and phase equilibria in binary van der Waals mixtures. Philosophical Transactions of the Royal Society. 1980;**298**:495-594. DOI: 10.1098/rsta.1980.0266

[10] Galicia-Luna LA, Richon D, Renon H. New loading technique for a vibrating tube densimeter and measurements of liquid densities up to 39.5 MPa for binary and ternary mixtures of the carbon dioxide-methanol-propane system. Journal of Chemical and Engineering Data. 1994; **39**(3):424-431. DOI: 10.1021:je00015a005

[11] Bouchot C, Richon D. Direct pressure–volume–temperature and vapor–liquid equilibrium measurements with a single equipment using a vibrating tube densimeter up to 393 K and 40 MPa: description of the original apparatus and new data. Industrial & Engineering Chemistry Research. 1998;**37**(8):3295-3304. DOI: 10.1021/ie970804w

[12] Wang S, Fauve R, Coquelet C, Valtz A, Houriez C, Artola PA, et al. Vapor–liquid equilibrium and molecular simulation data for carbon dioxide (CO_2)+ trans-1, 3, 3, 3-tetrafluoroprop-1-ene (R-1234ze (E)) mixture at temperatures from 283.32 to 353.02 K and pressures up to 7.6 MPa. International Journal of Refrigeration. 2019;**98**:362-371. DOI: 10.1016/j.ijrefrig.2018.10.032

[13] Meskel-Lesavre M, Richon D, Renon H. A new variable volume cell for determining vapor-liquid equilibria and saturated liquid molar volume by the static method. Industrial and Engineering Chemistry Fundamentals. 1981;**20**: 284-289. DOI: 10.1021:i100003a017

[14] Fontalba F, Richon D, Renon H. Simultaneous determination of PVT and

VLE data of binary mixtures up to 45 MPa and 433 K: A new apparatus without phase sampling and analysis. The Review of Scientific Instruments. 1984;**55**:944-951. DOI: 10.1063/1.1137870

[15] Tzirakis F, Stringari P, Von Solms N, Coquelet C, Kontogeorgis G. Hydrate equilibrium data for the CO_2+ N_2 system with the use of tetra-n-butylammonium bromide (TBAB), cyclopentane (CP) and their mixture. Fluid Phase Equilibria. 2016;**2016, 408**:240-247. DOI: 10.1016/j.fluid.2015.09.021

[16] Wagner W, Brachthäuser K, Kleinrahm R, Lösch HW. A new, accurate single-sinker densitometer for temperatures from 233 to 523 K at pressures up to 30 MPa. International Journal of Thermophysics. 1995;**16**(2): 399-411. DOI: 10.1007/BF01441906

[17] Bridgman PW. The volume of eighteen liquids as a function of pressure and temperature. Proceedings of the American Academy of Arts and Sciences. 1931;**66**(5):185-233. DOI: 10.2307/20026332

[18] Straty GC, Palavra AMF. Automated high temperature PVT apparatus with data for propane. Journal of Research of the National Bureau of Standards. 1984; **89**(5):375-383

[19] Goodwin ARH, Marsh KN, Wakeham WA. Measurement of the Thermodynamic Properties of Single Phases, Experimental Thermodynamics Vol VI IUPAC. Amsterdam: Elsevier; 2003

[20] Coquelet C, Ramjugernath D, Madani H, Valtz A, Naidoo P, Meniai AH. Experimental measurement of vapor pressures and densities of pure hexafluoropropylene. Journal of Chemical & Engineering Data. 2010;**55**(6): 2093-2099. DOI: 10.1021/je900596d

[21] Tanaka K, Higashi Y. Thermodynamic properties of HFO-1234yf (2, 3, 3, 3-tetrafluoropropene). International Journal of Refrigeration. 2010;**33**(3):474-479. DOI: 10.1016/j.ijrefrig.2009.10.003

[22] Moldover MR, Mehl JB, Greenspan M. Gas-filled spherical resonators: Theory and experiment. The Journal of the Acoustical Society of America. 1986; **79**(2):253-272. DOI: 10.1121/1.393566

[23] Trusler JPM, Zarari M. The speed of sound and derived thermodynamic properties of methane at temperatures between 275 and 375 K and pressures up to 10 MPa. The Journal of Chemical Thermodynamics. 1992;**24**(9): 973-991. DOI: 10.1016/S0021-9614(05) 80008-4

[24] Benedetto G, Gavioso RM, Spagnolo R, Marcarino P, Merlone A. Acoustic measurements of the thermodynamic temperature between the triple point of mercury and 380 K. Metrologia. 2004; **41**(1):74-98

[25] Ahmadi P, Chapoy A, Tohidi B. Density, speed of sound and derived thermodynamic properties of a synthetic natural gas. Journal of Natural Gas Science and Engineering. 2017;**40**: 249-266. DOI: 10.1016/j.jngse.2017.02. 009Get

[26] Al Ghafri SZ, Rowland D, Akhfash M, Arami-Niya A, Khamphasith M, Xiao X, et al. Thermodynamic properties of hydrofluoroolefin (R1234yf and R1234ze (E)) refrigerant mixtures: Density, vapour-liquid equilibrium, and heat capacity data and modelling. International Journal of Refrigeration. 2019;**98**:249-260. DOI: 10.1016/j. ijrefrig.2018.10.027

[27] Calvignac B, Rodier E, Letourneau JJ, Vitoux P, Aymonier C, Fages J. Development of an improved falling ball viscometer for high-pressure measurements with supercritical CO_2. The Journal of Supercritical Fluids. 2010;**55**(1):96-106. DOI: 10.1016/j. supflu.2010.07.012

[28] Daridon JL, Cassiède M, Paillol JH, Pauly J. Viscosity measurements of liquids under pressure by using the quartz crystal resonators. Review of Scientific Instruments. 2011;**82**(9):095114. DOI: 10.1063/1.3638465

[29] Caudwell D, Goodwin AR, Trusler JM. A robust vibrating wire viscometer for reservoir fluids: results for toluene and n-decane. Journal of Petroleum Science and Engineering. 2004;**44**(3–4): 333-340. DOI: 10.1016/j.petrol.2004.02.019

[30] El Abbadi J. Thermodynamic properties of new refrigerants [thesis]. Mines ParisTech; 2016

[31] Le N, Viscosité B. Définitions et dispositifs de mesure. Techniques de l'ingénieur. Constantes physico-chimiques. 2004:3K478

[32] Kashefi K, Chapoy A, Bell K, Tohidi B. Viscosity of binary and multicomponent hydrocarbon fluids at high pressure and high temperature conditions: Measurements and predictions. Journal of Petroleum Science and Engineering. 2013;**112**: 153-160. DOI: 10.1016/j.petrol.2013.10.021

[33] Tait RWF, Hills BA. Methods for determining liquid thermal conductivities. Industrial and Engineering Chemistry. 1964; **56**(7):29-35. DOI: 10.1021:ie50655a005

[34] Healy JJ, De Groot JJ, Kestin J. The theory of the hot-wire method for measuring thermal conductivity. Physica C. 1976;**82**(2):392-408. DOI: 10.1016/0378-4363(76)90203-5

[35] Marsh KN, Perkins RA, Ramires MLV. Measurement and correlation from 86 to 600K at pressures to 70 MPa. Journal of Chemical & Engineering Data. 2002;**47**:932-940. DOI: 10.1021/je010001m

[36] Chapman WG, Gubbins KE, Jackson G, Radosz M. SAFT: Equation-of-state solution model for associating fluids. Fluid Phase Equilibria. 1989;**52**: 31-38. DOI: 10.1016/0378-3812(89)80308-5

[37] Gil-Villegas A, Galindo A, Whitehead PJ, Mills SJ, Jackson G, Burgess AN. Statistical associating fluid theory for chain molecules with attractive potentials of variable range. The Journal of Chemical Physics. 1997;**106**(10):4168-4186. DOI: 10.1063/1.473101

[38] Gross J, Sadowski G. Perturbed-chain SAFT: An equation of state based on a perturbation theory for chain molecules. Industrial & Engineering Chemistry Research. 2001;**40**(4): 1244-1260. DOI: 10.1021/ie0003887

[39] Lafitte T, Apostolakou A, Avendano C, Galindo A, Adjiman CS, Müller EA, et al. Accurate statistical associating fluid theory for chain molecules formed from Mie segments. The Journal of Chemical Physics. 2013;**139**(15):154504. DOI: 10.1063/1.4819786

[40] Tillner-Roth R. Die thermodynamischen Eigenschaften von R 152a, R 134a and ihren Gemischen-Messungen Und Fundamentalgleichungen PhD. Germany: University of Hannover; 1993

[41] Mac Linden MO, Klein SA. A next generation refrigerant properties database. In: International Refrigeration and Air Conditioning conference 1996: paper 357. pp. 409-414

[42] Huber ML, Laesecke A, Perkins RA. Model for the viscosity and thermal conductivity of refrigerants, including a new correlation for the viscosity of R134a. Industrial & Engineering Chemistry Research. 2003;**42**: 3163-3178. DOI: 10.1021/ie0300880

[43] Klein SA, McLinden MO, Laesecke A. An improved extended

corresponding states method for estimation of viscosity of pure refrigerants and mixtures. International Journal of Refrigeration. 1997;**20**(3): 208-217. DOI: 10.1016/S0140-7007(96) 00073-4

[44] Poling E, Prausnitz JM. The Properties of Gases and Liquids. McGraw-Hill Professional. New York: Mcgraw-Hill; 2000

[45] Huber ML, Hanley HJM. The Corresponding-States Principle: Dense Fluids. Transport Properties of Fluids. Cambridge University Press, IUPAC; 1996. pp. 283-296

[46] Ely JF, Huber ML. A predictive extended corresponding states model for pure and mixed refrigerants including a new equation of state for R134a. International Journal of Refrigeration. 1994;**17**:18-31. DOI: 10.1016/0140-7007(94)90083-3

[47] Peng DY, Robinson DB. A new two-constant equation of state. Industrial and Engineering Chemistry Fundamentals. 1976;**15**(1):59-64. DOI: 10.1021:i160057a011

[48] Wong DSH, Sandler SI. A theoretically correct mixing rule for cubic equations of state. AICHE Journal. 1992;**38**(5):671-680. DOI: 10.1002: aic.690380505

[49] Renon H, Prausnitz JM. Local compositions in thermodynamic excess functions for liquid mixtures. AICHE Journal. 1968, 1968;**14**(1):135-144. DOI: 10.1002/aic.690140124

[50] Juntarachat N, Valtz A, Coquelet C, Privat R, Jaubert JN. Experimental mea surements and correlation of vapor-liquid equilibrium and critical data for the CO_2 + R1234yf and CO_2 + R1234ze (E) binary mixtures. International Journal of Refrigeration. 2014;**47**: 141-152. DOI: 10.1016/j.ijrefrig.2014. 09.001

[51] El Ahmar E, Valtz A, Paricaud P, Coquelet C, Abbas L, Rached W. Vapour–liquid equilibrium of binary systems containing pentafluorochemicals from 363 to 413 K: Measurement and modelling with Peng–Robinson and three SAFT-like equations of states. International Journal of Refrigeration. 2012;**35**(8):2297-2310. DOI: 10.1016/j.ijrefrig.2012.05.016

[52] Lemmon EW, Bell IH, Huber ML, McLinden MO. NIST standard reference database 23: Reference fluid thermodynamic and transport properties-REFPROP (National Institute of Standards and Technology, Boulder, CO) 2018: Version 10.0

[53] Coquelet C, Chareton A, Richon D. Vapour–liquid equilibrium measurements and correlation of the difluoromethane (R32) + propane (R290)+ 1, 1, 1, 2, 3, 3, 3-heptafluoropropane (R227ea) ternary mixture at temperatures from 269.85 to 328.35 K. Fluid Phase Equilibria. 2004; **218**(2):209-214. DOI: 10.1016/j. fluid.2003.12.009

[54] Soave G. Equilibrium constants from a modified Redlich-Kwong equation of state. Chemical Engineering Science. 1972;**27**(6):1197-1203. DOI: 10.1016/0009-2509(72)80096-4

[55] Michelsen ML. A modified Huron-Vidal mixing rule for cubic equations of state. Fluid Phase Equilibria. 1990;**60** (1–2):213-219. DOI: 10.1016/0378-3812 (90)85053-D

[56] Coquelet C, Valtz A, Théveneau P. Personal Communication and Confidential Data

[57] Laesecke A, Hafer RF. Viscosity of fluorinated propane isomers. 2. Measurements of three compounds and model comparisons. Journal of Chemical & Engineering Data. 1998;**43**(1):84-92. DOI: 10.1021/je970186q

[58] Marrucho IM, Oliveira NS, Dohrn R. Vapor-phase thermal conductivity, vapor pressure, and liquid density of R365mfc. Journal of Chemical & Engineering Data. 2002;**47**(3):554-558. DOI: 10.1021/je015534

[59] Wang D, Ling X, Peng H, Liu L, Tao L. Efficiency and optimal performance evaluation of organic Rankine cycle for low grade waste heat power generation. Energy. 2013;**50**:343-352. DOI: 10.1016/j.energy.2012.11.010

Section 2

Applications

Chapter 3

Waste Heat Recovery from Fossil-Fired Power Plants by Organic Rankine Cycles

Qiang Liu

Abstract

More than 60% of the world's electricity is still produced from fossil-fired power plants. Recovering heat from flue gas, drained water, and exhaust steam which are discharged in power plants by organic Rankine cycles (ORCs) to generate power is an efficient approach to reduce fossil fuel consumption and greenhouse gas emissions. This chapter proposes conceptual ORC systems for heat recovery of drain from continuous blowdown systems, exhaust flue gas from boilers, and exhaust steam from turbines. The waste heat source temperatures range from 30 to 200°C. Environmentally friendly and nonflammable working fluids including R134a, R1234ze, R236ea, R245fa, R1233zd, and R123 were selected as the working fluids. The parameters of ORC systems were optimized, and the thermodynamic performance was analyzed. The suitable ORC layouts for various kinds of heat sources including drained water, flue gas, and steam were discussed with selecting the proper working fluids. The gross power output of a coal-fired power plant can be increased up to 0.4% by an ORC using the waste heat from the boiler flue gas. The ORCs using turbine exhaust steam with the cooling water as low as 5°C can generate 2–3% more power for a power unit.

Keywords: coal-fired power plant, waste heat recovery, organic Rankine cycle, parametric optimization, working fluid, thermodynamic analysis

1. Introduction

In 2018, the total world electricity generation reached 26,672 TWh [1]. About 64.15% of the total world electricity generation is still from fossil fuels (coal, natural gas, and oil) [1], especially in China, the USA, Japan, Russia, and India. China accounted for 26.2% of the world electricity generation [2]. The installed fossil-fired power capacity is now increased to 1143.67 GW in China which accounts for about 60% of the total installed power capacity; however, 70.4% of electricity was generated from fossil fuels [2]. Now, China is also the world's greatest emitter of greenhouse gases from the burning of fossil fuels.

The development of large-capacity and high-temperature ultra-supercritical fossil-fired power unit is the trend in the world. However, a large number of waste heat are discharge in a power plant. **Figure 1** shows waste heat sources in a typical coal-fired power unit. Therefore, recovering the waste heat is a main approach to

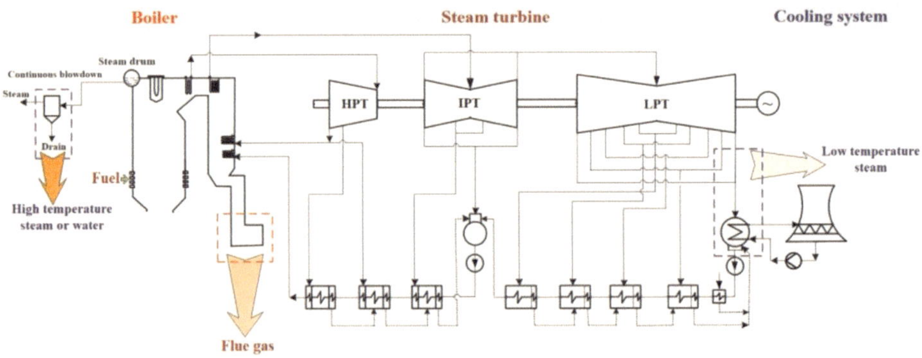

Figure 1.
Schematic of the waste heat sources can be recovered by ORCs in a typical coal-fired power unit.

further improve thermal efficiencies and reduce greenhouse gas emissions for fossil-fired power plants [3].

Continuous blowdown systems are used to purify the steam by removing suspended solids in drum boilers for subcritical power plants as shown in **Figure 1**. A part of the boiler water with temperatures up to 360°C from the drum is discharge to the flash tank. In the flash tank, some of the water flashes to the steam, and then the steam enters into a deaerator or a feedwater heater for feedwater preheating, while the drain is generally discharge to the sewer. The drain temperature is higher than 150°C; however, the heat in the drained water is lost through direct discharge. The waste heat in the drained water can be recovered by an organic Rankine cycle to generate electric power [4].

A modern power station boiler produces large quantities of flue gas. The exhaust flue gas temperatures generally range from 120–140°C for power station boilers. Heat loss due to the exhaust flue gas is the largest heat loss in a power station boiler which significantly affects the boiler thermal efficiency. For large- and medium-capacity boilers, the exhaust flue gas heat loss accounts for 4–8% of the total boiler heat input from fossil burning [3]. In actual operation, the flue gas temperature may be higher than the design value which results in a lower boiler thermal efficiency and a higher fuel consumption. The heat in the exhaust flue gas can be used to preheat feedwater by a low-pressure economizer [5–7] and pre-dry brown coal [8–11]. The waste heat of exhaust flue gas can also be used to drive an organic Rankine cycle to produce electric power [5, 12–14].

The latent heat of the exhaust steam from a condensing turbine is released to the cooling water at the condenser as shown in **Figure 1**. Even for a modern steam turbine, 50–60% of the heat input is discharge to the heat sink. Seawater is generally used as the cooling water for the offshore power plants. The seawater at high-latitude regions or at deep sea (depth down to 1000 m) has a temperature as low as 5°C. However, the condensing turbine exhaust pressures are generally higher than 3.5kPa with condensation temperatures not less than 26.7°C due to the limits from the wetness loss, last-stage blade length, and droplet erosion. Thus, the temperature difference between the exhaust steam and seawater can be used to drive an ORC to produce electricity. An ORC can also use a part of the extracted steam from low-pressure turbine and boiler flue gas [14] or only the coldest extraction steam [15] to generate additional electricity.

In coal-fired power plants, there are different kinds of heat sources which can be utilized by ORCs. The waste heat sources involve different heat transfer fluids (gas, water, and steam) and large scales of capacity and temperature. The exhaust flue gas with temperatures up to 140°C from a supercritical boiler which contains

10–30 MW heat can be recovered. The turbine exhaust steam at 30°C contains more than 300 MW heat for a large-capacity power unit. Therefore, this chapter focuses on the thermodynamic performance of ORCs using the waste heat from coal-fired power plants. The suitable cycle layouts are discussed with considering the waste heat source characteristics. The evaporation temperature and pressure of the working fluid will be optimized to maximize the cycle net power output. The proper working fluid is selected according to the thermodynamic performance. The power generation potentials are also evaluated for various kinds of waste heat sources.

2. ORC model

The ORCs using low-boiling organic working fluids are the unequaled technology for producing electricity from medium-low-temperature heat sources [16]. The ORCs have distinctive features of being simple layouts, being efficient for off-design conditions, and being reasonable cost-effectiveness. Therefore, ORC systems have been widely used to convert geothermal, solar, biomass, waste heat, and ocean thermal energy to electricity. The theoretical basis for ORCs has been described elsewhere [16–22]. This section briefly describes the cycle layouts and working fluids selection.

2.1 Cycle layouts

The ORCs can be classified into subcritical and supercritical cycles according to the working fluid evaporation pressure. Considering the saturated vapor line shapes of working fluids, the subcritical ORC with saturated vapor at the turbine inlet is called o2 cycle, and that with superheated vapor at the turbine inlet is called o3 cycle which uses a dry working fluid or b3 cycle which uses a wet working fluid, and the supercritical ORC is called s2 cycle [17]. The cycle layouts are selected according to the heat source characteristics and the working fluid critical temperature. **Figure 2(a)** shows a typical ORC system with a wet cooling system with the *T-s* diagrams for o2, o3, b3, and s2 cycles shown in **Figure 3**. The working fluid is heated to a saturated or a superheated vapor from a subcooled liquid by the heat source fluid in heat exchangers. Thus, the energy balance for working fluid evaporation can be uniformly expressed as

$$\dot{m}_{\mathrm{H}}(h_{\mathrm{Hin}} - h_{\mathrm{Hout}}) = \dot{m}_{\mathrm{O}}(h_{\mathrm{O}} - h_3) \tag{1}$$

where h is the specific enthalpy, the subscript H represents the heat source, and the subscript O represents the organic fluid.

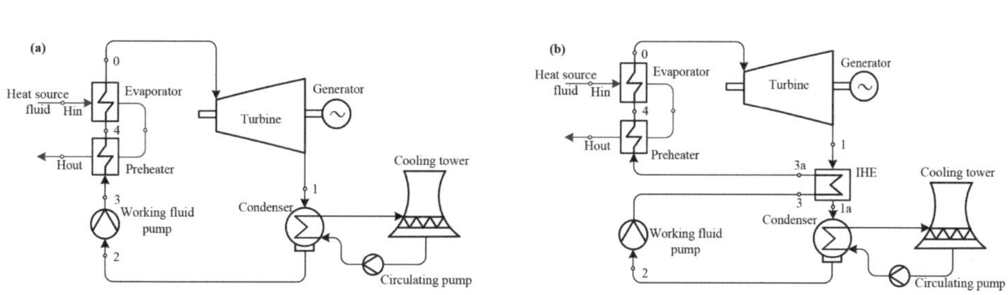

Figure 2.
Schematics of (a) a typical ORC and (b) an ORC with an IHE.

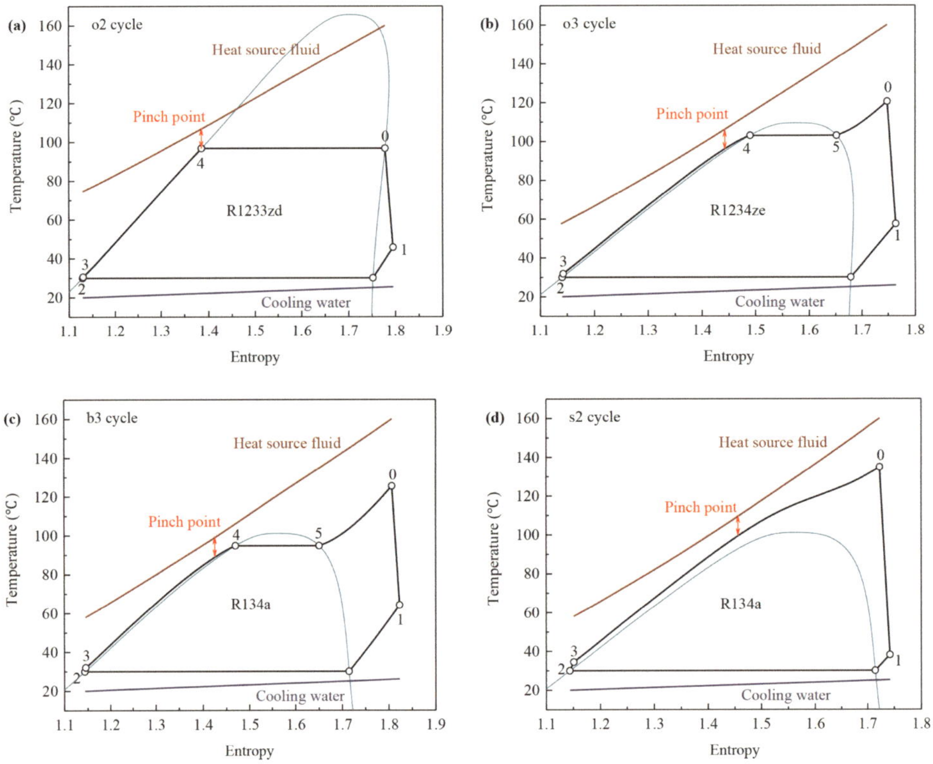

Figure 3.
Temperature-entropy diagrams for basic ORCs: (a) o2 cycle, (b) o3 cycle, (c) b3 cycle, and (d) s2 cycle.

The power generated by the turbine was

$$\dot{W}_{\mathrm{T}} = \dot{m}_{\mathrm{O}}(h_0 - h_1) \tag{2}$$

The power consumed by the working fluid feed pump was

$$\dot{W}_{\mathrm{FP}} = \dot{m}_{\mathrm{O}}(h_3 - h_2) \tag{3}$$

The net power output by the ORC system was defined as the turbine power output which subtracts the parasitic power consumed by the working fluid feed pump:

$$\dot{W}_{\mathrm{net}} = \dot{W}_{\mathrm{T}}\eta_{\mathrm{m}}\eta_{\mathrm{g}} - \dot{W}_{\mathrm{FP}} \tag{4}$$

The cycle thermal efficiency is defined as

$$\eta_{\mathrm{th}} = \frac{\dot{W}_{\mathrm{net}}}{\dot{Q}_{\mathrm{in}}} \tag{5}$$

The turbine exhaust vapor temperature could still be relative high; the superheated vapor could be worth preheating the working fluid in an internal heat exchanger (IHE), also referred to as a recuperator or regenerator, before the preheater [16–18] as shown in **Figure 2(b)**. Eq. (1) for an ORC with an IHE can be rewritten as

$$\dot{m}_{\mathrm{H}}(h_{\mathrm{Hin}} - h_{\mathrm{Hout}}) = \dot{m}_{\mathrm{O}}(h_0 - h_{3\mathrm{a}}) \tag{6}$$

The use of an IHE can improve the preheater inlet temperature by the turbine exhaust vapor. For the heat source without the limitation of the outlet temperature [18, 23], the heat source fluid outlet temperature increases because the working fluid is preheated as shown in **Figure 4(a)** and the cycle net power output will maintain constant or increase slightly [24]. For the heat source with a limited outlet temperature as shown in **Figure 4(b)**, the increase in the preheater inlet temperature results in an increase in the working fluid mass flow rate or the optimal evaporation temperature; thus, the net power output increases.

2.2 Systematic parameters

The matching characteristics between the heat source fluid and the working fluid affect the ORC net power output. In subsequent sections, the working fluid evaporation temperature or pressure will be optimized to maximize the cycle net power output. The operating parameters and boundary conditions for the ORCs are listed in **Table 1**. The parametric optimization and thermodynamic analyses are based on the parameters listed in **Table 1** except the ORC in Section 5.2. The heat losses and pressure drops are neglected except instructions.

The pinch point temperature difference is an important parameter for system design and optimization [25, 26]. During the working fluid evaporation, the pinch point may locate at the bubble point, a subcooled liquid state near the critical point and a superheated state [26] which depends on the heat source temperature drop

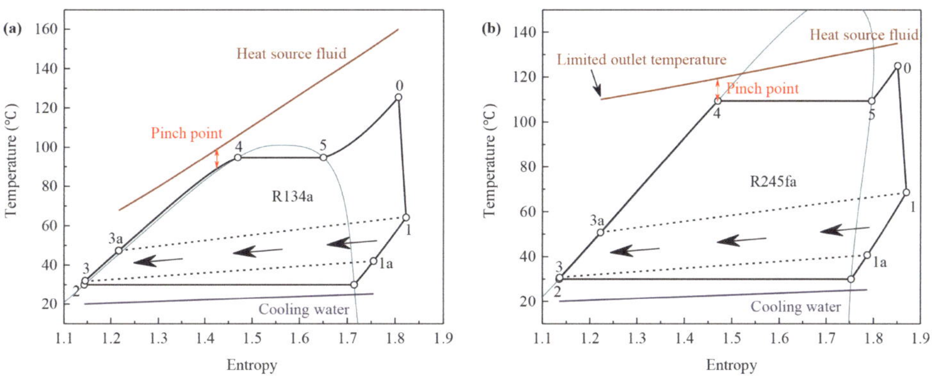

Figure 4.
Temperature-entropy diagrams of ORCs with an IHE for heat source fluids: (a) without outlet temperature limitation and (b) with limited outlet temperature.

Parameters	Values	Unit
ORC turbine isentropic efficiency	85	%
Working fluid pump efficiency	80	%
Generator efficiency	98	%
Turbine mechanical efficiency	98	%
Evaporator pinch point temperature difference	10	°C
IHE pinch point temperature difference	10	°C
Condenser pinch point temperature difference	5	°C
Condensation temperature	30	°C

Table 1.
Parameters and boundary conditions for the ORC systems.

Working fluid	T_{cr}/°C	p_{cr}/MPa	Safety class	ODP	GWP
R134a	101.06	4.059	A1	0	1430
R1234ze(E)	109.36	3.635	A2L	0	6
R236ea	139.29	3.420	n.a	0	1410
R245fa	154.01	3.651	B1	0	1030
R1233zd(E)	165.60	3.572	A1	0.003	1
R123	183.68	3.662	B1	0.01	77

Table 2.
Physical and environmental properties of the six working fluids [29].

and heat capacity and the working fluid thermophysical properties. Thus, each heat exchanger is divided into 100 sections with equal heat-flow interval [25, 27] for determining the pinch point during the parameter optimization.

2.3 Working fluids

Considering the operating safety, nonflammable working fluids are selected for the ORCs using the waste heat in coal-fired power plants. Moreover, the working fluids should be environmentally friendly with a lower ozone depletion potential (ODP) and global working potential (GWP). The critical temperature has been regarded as a criterion for working fluid selection [17, 28]; thus, the studied fluids should cover a wide range of critical temperatures. The selected working fluids here are R134a (1,1,1,2-tetrafluoroethane), R1234ze (1,3,3,3-tetrafluoropropene), R236ea (1,1,1,2,3,3-hexafluoropropane), R245fa (1,1,1,3,3-pentafluoropropane), R1233zd (trans-1-chloro-3,3,3-trifluoropropene), and R123 (2,2-dichloro-1,1,1-trifluoroethane). The physical and environmental properties of the selected working fluids are listed in **Table 2**. The thermophysical properties for the working fluids were calculated using REFPROP 9.1 [29].

3. Waste heat recovery from a boiler blowdown system

3.1 System description

A continuous blowdown system is generally used to purify the steam to maintain acceptable levels of total dissolved solids for a subcritical drum boiler. Part of the saturated boiler water at a higher salt concentration is discharge from the drum to the flash tank [3, 30]. The pressure decreases as the saturated boiler water enters the flash tank. The excess energy in the boiler water is given up as some of the saturated water flashes to the saturated steam. The steam then enters into the deaerator to preheat the boiler feedwater, and the drain is generally discharge to the sewer. The energy in the drained liquid is lost through direct discharge.

The drain temperature is generally higher than 150°C. Here, an organic Rankine cycle (ORC) is designed to recover the waste heat of the drained water from a continuous blowdown system for power generation and then to improve the overall thermal efficiency of the power plant [4]. The organic working fluid is heated by the waste heat in the discharged drain and then generates power by expansion through a turbine as shown in **Figure 5**.

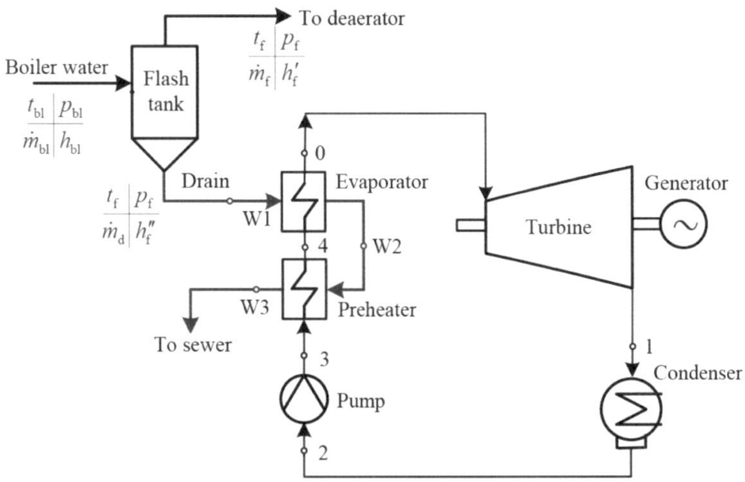

Figure 5.
Schematic of an ORC power generation system using the blowdown waste heat from a subcritical boiler [4].

The energy balance for the flash tank can be expressed as

$$\dot{m}_{\rm bl}h_{\rm bl}\eta_{\rm f} = \dot{m}_{\rm f}h''_{\rm f} + \dot{m}_{\rm d}h'_{\rm f} \tag{7}$$

where $\dot{m}_{\rm bl}$ is the boiler water flow rate, $h_{\rm bl}$ is the specific enthalpy of the boiler water, $\eta_{\rm f}$ is the flash tank efficiency which is set to be 0.98, $\dot{m}_{\rm f}$ is the saturated steam flow rate, $\dot{m}_{\rm d}$ is the drained water flow rate, $h''_{\rm f}$ is the specific enthalpy of the saturated steam, and $h'_{\rm f}$ is the specific enthalpy of the drained water.

The mass balance for the flash tank can be expressed as

$$\dot{m}_{\rm bl} = \dot{m}_{\rm f} + \dot{m}_{\rm d} \tag{8}$$

The flow rate of the saturated steam flashed from the boiler water increases as the flash tank pressure, $p_{\rm f}$, decreases. The flash tank pressure generally is set to be the deaerator pressure which ranges from 0.8 to 1.1 MPa for subcritical steam turbines. Thus, 40–50% of the boiler water is flashed into saturated steam.

The main-steam flow rates of typical subcritical 600 MW power units range from 1600 to 2000 t/h for the BRL (boiler rated load) condition. About 13–16 t/h boiler water enters the flash tank if the blowdown rate is 0.8% which means that the drained water flow rate can reach 6–10 t/h. A power plant generally has 2–8 power units; thus, the drain water from the continuous blown down systems can be collected, and the flow rate may be reached 80 t/h.

A typical subcritical 600 MW boiler is analyzed in this section. The boiler parameters for the BRL condition are listed in **Table 3**. The boiler water flow rate in a blowdown system is set to be 17.1 t/h. At the flash tank outlet, the drain flow rate is 9 t/h according to Eqs. (7) and (8). Assume a power plant has same four power units; thus, the total drain flow rate is 10 kg/s (36 t/h). The collected drain is used to drive an ORC as shown in **Figure 5**. Considering the heat loss, the drain temperature to the ORC system is set to be 180°C.

3.2 Thermodynamic performance

The turbine inlet temperatures (pressure) were optimized to maximize the cycle net power output for the ORCs with saturated vapor at the turbine inlet (o2 cycle)

Parameters	Values	Unit
Main-steam flow rate	2011	t/h
Steam drum pressure	18.5	MPa
Boiler water temperature in drum	359.3	°C
Flash tank pressure	1.03	MPa
Drain temperature	181.2	°C
Flash tank efficiency	98	%
Drain flow rate	10	kg/s
Drain pressure	1.03	MPa
Drain temperature at the evaporator inlet	180	°C

Table 3.
Blowdown system parameters of a 600 MW boiler for the BRL condition.

Working fluid	Cycle	$T_{T,in}$/°C	$P_{T,in}$/MPa	Power output/kW	Heat utilization ratio/%
R245fa	o2	116.3	1.79	660	70.99
R1233zd(E)	o2	109.8	1.28	622	68.08
R123	o2	106.4	0.90	599	65.88
R134a	b1	169.6	3.06	548	75.81
R1234ze	o3	153.0	2.63	567	76.47
R236ea	o3	121.4	2.06	677	77.81
R134a	S2	163.4	8.57	749	77.10
R1234ze	S2	163.7	7.48	746	77.10
R236ea	S2	160.7	4.95	764	77.10

Table 4.
Parameters of the ORC for waste heat recovery from a blowdown system.

using R245fa, R1233zd(E), and R123. The optimized turbine inlet temperatures and the maximum net cycle power outputs are listed in **Table 4**. The results show that the optimal turbine inlet temperature decreases as the critical temperature of the working fluid increases. The net power output as well as the waste heat utilization rate decreases for working fluids with higher critical temperatures. The net cycle power output is 660 kW for the o2 cycle using R245fa, while the net cycle power output is only 599 kW for the o2 cycle with R123.

Here, the superheated ORC uses a working fluid with a lower critical temperature (R134a, R1234ze, and R236ea) for recovering the waste heat from the drained water. The turbine inlet temperature and pressure are optimized simultaneously to obtain the maximum cycle net power output using the generalized reduced gradient (GRG) method which has been successfully used in previous work [4, 27]. Both the optimized turbine inlet temperature and pressure are higher for the working fluid with lower critical temperature as shown in **Table 4**. However, the working fluid with higher critical temperature generates a higher net power due to the higher working fluid flow rate. The net cycle power output with R236ea is higher than that with R134a or R1234ze.

A supercritical ORC provides a better match between the working fluid and the heat source fluid temperature profiles. The working fluids with much lower critical temperatures, such as R134a, R1234ze, and R236ea, are considered here for supercritical ORCs. The turbine inlet temperature and pressure are optimized simultaneously to maximize the cycle net power output considering the pinch point temperature difference limit. Compared with subcritical ORCs, a supercritical ORC produces 13–37% more net cycle power as shown in **Table 4**. The optimal turbine inlet temperatures for supercritical ORCs with the selected working fluids are very close; however, the working fluid with a lower critical temperature has a higher optimal turbine inlet pressure. The high operating pressure and heat transfer deterioration due to the large specific heat near the critical point must be considered in the system design [4]. The supercritical ORC using R236ea generates the highest net power (764 kW) among the considered ORCs with the selected working fluids.

4. Waste heat recovery from boiler exhaust flue gas

The heat loss in the exhaust flue gas is the largest heat loss in a power station boiler which typically accounts 70–80% of the total boiler heat losses [30]. The temperature of the exhaust flue gas from a power station boiler generally ranges from 120–140°C. Unfortunately, the flue gas temperature may be higher than the design value in operation which results in more heat losses and lower boiler thermal efficiencies. The exhaust flue gas can be cooled by a low-boiling organic working fluid with the fluid then passing through a turbine to generate power. The potential decrease in the exhaust flue gas temperature is 10–30°C. Supercritical 1000 MW boilers for power plants consume about 350 tons of bituminous coal and generate about 3×10^6 Nm^3 flue gas per hour for the BRL condition. More than 10 MW waste heat can be recovered to drive an ORC when the flue gas temperature is reduced by 10°C. This section discusses the thermodynamic performance of ORCs using the flue gas waste heat from a supercritical boiler.

4.1 Coal-fired boiler

More than 90 supercritical 1000 MW coal-fired power units are being operated in China with more 1000 MW power units planned to replace many current small-capacity and low-efficiency power units [31]. Therefore, a modern supercritical 1000 MW power unit is taken here as a case study. The boiler operating parameters for the BRL condition are listed in **Table 5**. The main-steam temperature is 605°C with a flow rate of 800.5kg/s. The exhaust steam from the high-pressure steam turbine is then reheated to 603°C in the reheater with a flow rate of 649kg/s. The feedwater temperature at the economizer inlet is 299.4°C with no water and steam assumed to be lost in the boiler. The exhaust flue gas temperature is 135°C, and the boiler thermal efficiency is 92.57%.

The ultimate analysis (UA) of the fuel is useful for the calculation of air and flue gas quantities and other combustion calculations [3, 30]. The coal consists of five elements (carbon, hydrogen, nitrogen, oxygen, and sulfur), moisture, and ash. A typical bituminous coal is used here as the boiler fuel. The elemental weights of the coal are determined on as-received basis: C_{ar} = 61.7%, H_{ar} = 3.67%, O_{ar} = 8.56%, N_{ar} = 1.12%, S_{ar} = 0.6%, A_{ar} = 8.8%, and M_{ar} = 15.55%. The lower calorific value (LCV) of the bituminous coal is 23.47 MJ/kg. The boiler consumes 353.5 tons of bituminous coal and generates 3.055×10^6 Nm^3 flue gas per hour for the BRL condition.

Sulfur in the coal on combustion forms SO_2 and partly SO_3. Sulfuric acid will be formed because of the reaction between SO_3 and H_2O in the flue gas at lower temperatures, which then condenses on the tube surface below the acid dew point and leads to the corrosion of the tubes [3, 30]. Therefore, the flue gas was assumed to be cooled to 110°C by the organic working fluids to avoid low-temperature corrosion.

Parameters	Values	Unit
Main-steam temperature at superheater outlet	605	°C
Main-steam pressure at superheater outlet	26.13	MPa
Main-steam flow rate	800.49	kg/s
Steam temperature at reheater outlet	603	°C
Reheated steam pressure at reheater outlet	4.57	MPa
Steam temperature at reheater inlet	347.7	°C
Reheated steam pressure at reheater inlet	4.80	MPa
Reheated steam flow rate	648.96	kg/s
Feedwater temperature	299.4	°C
Boiler thermal efficiency	92.57	%
Exhaust flue gas temperature	135	°C
Lower calorific value (LCV) of coal	23469.7	kJ/kg

Table 5.
Operating parameters of a supercritical 1000 MW boiler for the BRL condition.

4.2 ORC system driven by waste heat from boiler flue gas

Figure 6 shows an ORC system using waste heat of flue gas from a boiler for power generation. A counterflow heat exchanger is used here to recover the waste heat because the flue gas temperature is much lower. The heat exchanger tubes are placed in an in-line arrangement for a lower flue gas pressure drop. The flue gas longitudinal flows outside the tubes. The working fluid is heated to a vapor from a subcooled liquid in the tubes of the heat exchanger.

The heat source temperature is a key parameter which influences the choice of cycle types and working fluids [16–19, 23]. The flue gas temperature should be higher than the acidic dew point to avoid low-temperature external corrosion; thus, the temperature drop of the exhaust flue gas from power station boiler is much lower. Considering the lower temperature drop and the higher flue gas temperature at the preheater outlet, an internal heat exchanger (IHE) is used in the ORC system to improve the working fluid temperature at the preheater inlet by the turbine exhaust vapor and then decrease the temperature difference between the flue gas and the working fluid.

The energy balance in the heat exchanger can be expressed as

$$\dot{B}V_{\mathrm{fg}}\left(I_{\mathrm{fg}}^{\mathrm{in}} - I_{\mathrm{fg}}^{\mathrm{out}}\right)\eta_{\mathrm{HE}} = \dot{m}_{\mathrm{O}}(h_{\mathrm{O}} - h_{3\mathrm{a}}) \tag{9}$$

where $\dot{B}$ is the coal consumption rate, V_{fg} is the specific volume flue gas based on 1 kg coal (unit flue gas volume), $I_{\mathrm{fg}}^{\mathrm{in}}$ is the flue gas enthalpy at the heat exchanger inlet, and $I_{\mathrm{fg}}^{\mathrm{out}}$ is the flue gas enthalpy at the heat exchanger outlet, η_{HE} is the heat exchanger efficiency which considering the heat loss, $\dot{m}_{\mathrm{o}}$ is the organic working fluid flow rate, h_{O} is the working fluid specific enthalpy at the heat exchanger outlet, and $h_{3\mathrm{a}}$ is the specific enthalpy at the IHE outlet.

The unit gas weights (volumes) for the BRL condition were calculated using the ultimate analysis [3, 30]. The excess air coefficient in the flue gas was assumed to be 1.3. The heat-flow rate is 29.26 MW when the flue gas temperature is reduced from 135 to 110°C. The total heat loss is assumed to be 5% during the heat transfer due to the radiation and convection; thus, 27.8 MW heat is transferred to the organic

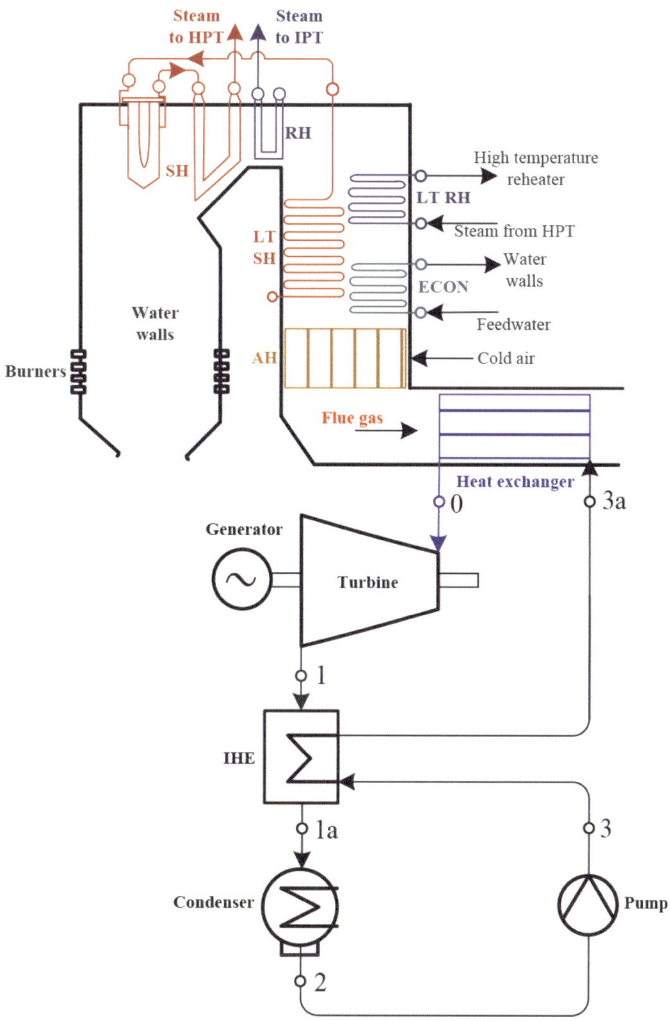

Figure 6.
Schematic of an ORC system using waste heat of flue gas from a boiler.

working fluids. The heat losses and pressure drops in the ORC system are neglected to simplify the calculations.

4.3 Thermodynamic performance

Both the o2 cycle and the o3 cycle are studied with the working fluids of R123, R1233zd, R245fa, and R236ea. The evaporation temperatures (pressures) of the selected working fluids were optimized for the o2 cycle to maximize the net cycle power output. The optimized parameters and system performance are listed in **Table 6**. **Figure 7(a)** shows the T-Q diagrams of the selected working fluids for o2 cycle. The optimal evaporation temperature of the o2 cycle is lower for the working fluid with a higher critical temperature which results in a higher heat load for the isothermal evaporation process and a better match between the working fluid and the flue gas temperatures as shown in **Figure 7(a)**. The working fluid with a higher critical temperature also has a lower evaporation pressure which leads to a lower working fluid pump power consumption. For example, 61.7% of the total heat flow is used for R123 boiling process, while 54.3% of the total heat flow is needed for R236fa preheating which process has a large temperature difference. Therefore, the

Working fluid	Cycle	$T_{T,in}$/°C	$P_{T,in}$/MPa	Power out/kW	Thermal efficiency/%
R236ea	o2	113.67	2.09	3783.0	13.61
R245fa	o2	111.84	1.63	3845.9	13.84
R1233zd(E)	o2	110.69	1.31	3876.9	13.95
R123	o2	109.60	0.97	3937.7	14.17
R134a	b3	125	3.85	3502.0	12.60
R1234ze	o3	125	2.74	3570.6	12.85
R236ea	o3	125	1.97	4025.5	14.48
R245fa	o3	125	1.55	4070.6	14.65
R1233zd(E)	o3	125	1.25	4074.6	14.66
R123	o3	125	0.93	4118.6	14.82
R134a	s2	125	4.26	3484.8	12.54
R1234ze	s2	125	3.82	3576.5	12.87

Table 6.
Comparison of ORCs using flue gas waste heat from a 1000 MW boiler.

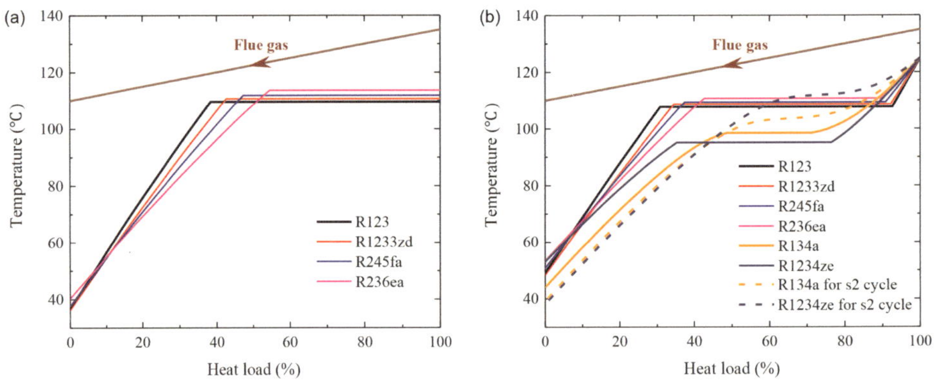

Figure 7.
T-Q diagrams for (a) the o2 cycles and (b) the cycles with superheated vapor or supercritical fluid at the turbine inlet.

o2 cycle using a working fluid with a higher critical temperature produces more net power as shown in **Table 6**. The o2 cycle using R123 generates about 4% more net power than that using R236ea.

Increasing the heat load of the isothermal evaporation process can better match the temperature profiles between the flue gas and the working fluid due to the lower temperature drop of the flue gas. The subcritical cycle using superheating process can further improve the temperature matching between the flue gas and the working fluid. The turbine inlet temperatures of the selected working fluids were set here to be 125°C for the subcritical cycles, and then the evaporation pressures were optimized to obtain the maximum net cycle power with the minimum temperature difference between the flue gas and the working fluid not less than 10°C. Compared to the o2 cycle, the use of superheating results in a slight decrease in the evaporation pressure and then reduces the power consumption of the working fluid pump. The turbine exhaust vapor temperature increases about 18°C due to the superheat degree at the turbine inlet; however, the working fluid temperature in the IHE can be increased by 12–13°C by the waste heat from the turbine exhaust vapor. Therefore, the average heat transfer temperature difference between the flue gas and the working fluid decreases, and the o3 cycle produces more power than the o2 cycle as shown in **Table 6**.

In this case study, the working fluid with a higher critical temperature needs a lower heat load for superheating and a higher heat load for boiling as shown in **Figure 7(b)**. For instance, the heat load for R123 superheating accounts for 7.4%, and the heat load for R123 boiling accounts for 61.8% of the total heat load; however, the heat load for R134a superheating accounts for 28.8%, and the heat load for R134a boiling only accounts for 22.4% of the total heat load because the boiling temperature is close to its critical temperature. The working fluids with lower critical temperatures have higher optimal evaporation pressures which also results in more power consumed by the working fluid pump. The ORC with superheating using R123 as the working fluid can generate 17.6% more power than that using R134a.

The supercritical ORCs using R1234ze and R134a which has a lower critical temperature are analyzed and compared here. The turbine inlet temperature is also set to be 125°C with the turbine inlet pressure of 1.05 p_{cr} because the turbine inlet temperature is very close to the critical temperature. The supercritical ORCs with R134a and R1234ze give a better matching of the temperature profiles between the working fluid and the flue gas compared to the subcritical ORCs as shown in **Figure 7(b)** which results in a higher turbine power output; however, the power consumed by the working fluid pump increases with increasing evaporation pressure which offsets the advantages of the supercritical ORCs. For example, the net power output of the supercritical ORC with R134a is even lower than the subcritical ORC for this case study. The subcritical ORC with superheating (o3 cycle) shows a better thermodynamic performance for lower temperature drop heat sources. The maximum power is produced by o3 cycle with R123 for waste heat recovery from the flue gas at a temperature of 135 °C, followed by o3 cycle with R1233zd and R245fa in this case study.

The volume flow rate at the turbine inlet as well as the turbine outlet/inlet volumetric flow ratio is important for the turbine design [17, 26]. **Figure 8(a)** shows the thermal efficiencies as a function of the turbine inlet volume flow rates of the working fluids for the optimal conditions. The mass flow rates of the selected working fluids are very close which range from 119 to 146 kg/s. However, a lower turbine inlet pressure leads to a lower density; thus, the working fluids evaporated at lower pressures have a much higher volume flow rates. The volume flow rate of R123 is the highest for the o2 cycle among the working fluids. Compared to the o2

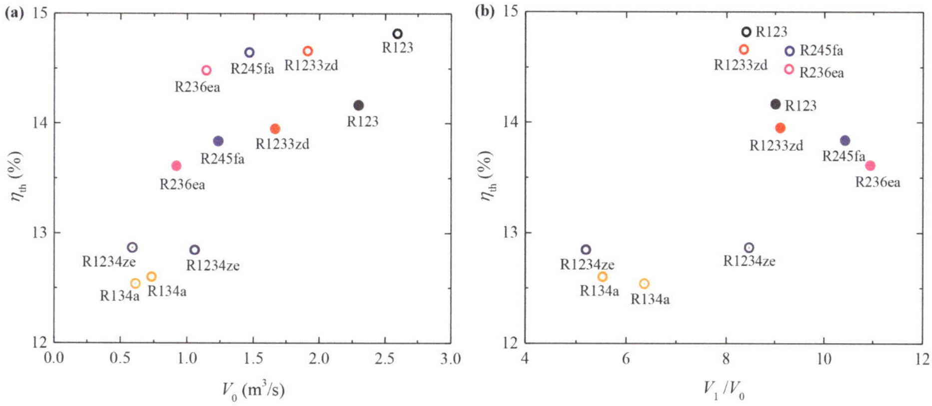

Figure 8.
Thermal efficiencies versus (a) volume flow rate, V_0, at the turbine inlet and (b) turbine outlet/inlet volumetric flow rate ratios for the optimal condition: ● saturated vapor, ○ superheated vapor, and ⊙ supercritical fluid.

cycle, the volume flow rate is increased by 10–20% for the o3 cycle due to the lower turbine inlet pressure as shown in **Figure 8(a)**, while the thermal efficiency can be increased by 4.6–6.4%. Both the R134a and R1234ze have lower volume flow rates with lower thermal efficiencies. The working fluid has a higher volume flow rate at the turbine inlet, also somewhat has a higher thermal efficiency as shown in **Figure 8(a)**.

Figure 8(b) shows the thermal efficiencies as a function of the VR (turbine outlet/inlet volumetric flow rate ratios, V_1/V_0) at the optimum turbine parameters for the maximum power output. Compared to the o2 cycle, the expansion ratio in the turbine decreases for the o3 cycle with the same condensation pressure which also leads to a decrease in the VR. Both the expansion ratios of R134a and R1234ze are much lower in the subcritical ORC because of the higher condensation pressures which also results in lower VRs. The VRs of R134a and R1234ze are increased in the supercritical ORC. Among the selected working fluids, R123, R1233zd, and R245fa have high thermal efficiencies with high-turbine-inlet volume flow rates and turbine outlet/inlet volumetric flow rate ratios. R1233zd is considered to be the most suitable working fluid for the ORC which recovers waste heat from the boiler flue gas with taking the safety and environment protection into account.

5. Heat recovery from turbine exhaust steam

5.1 ORCs driven by exhaust steam from a back-pressure turbine

Back-pressure steam turbines supply not only electricity but also the steam and heat for processes. The exhaust steam of the back-pressure steam turbine is directly used to supply heat or steam to the facilities without condensation. Ideally, the exhaust steam and the electricity from the back-pressure steam turbine are supplied to the same users [32]. The back-pressure steam turbine does not have any cold source loss (heat loss released directly to the environment). Therefore, the back-pressure steam turbines are efficient and have been widely used in industrial applications such as oil refineries, petrochemical plants, and cogeneration [33].

The exhaust steam pressure generally is set to be the demand pressure from the facility or outside needs [33]; thus, a lower expansion ratio results in a lower enthalpy drop and small power output in a back-pressure steam turbine. Only fewer turbine stages are used due to the lower enthalpy drop which leads to a simple structure and a lower cost for a back-pressure steam turbine.

Process steam/heat demand and electricity demand change independently according to the season or facility production demand [33]. An imbalance between process steam/heat and electricity demands is one of the most common problems in actual operation. The steam/heat demand is the primary requirement with electricity demand a secondary consideration to solve the imbalance demands because the steam and heat cannot economically and conveniently be transported over a long distance. Adjusting the main-steam flow rate of a back-pressure turbine is the major solution to meet the steam/heat demand. Thus, the back-pressure turbine power (electricity) output varies with the steam/heat demand. However, the thermal efficiency of the back-pressure steam turbine is significantly decreased with decreasing electricity output at partial loads due to the lower enthalpy drop. For example, the turbine isentropic efficiency (relative internal efficiency) can be decreased from 85 to 60% for lower steam flow rates which leads to an increase in the heat consumption of power generation [34].

The imbalance between steam/heat and electricity demands reduces the economic performance of back-pressure steam turbines. Part of this risk will be mitigated if the back-pressure steam turbine can be operated at the optimum condition (design condition) and the excess steam which beyond the steam/heat demand is used to generate electricity by an ORC which exports electricity directly into the grid [32]. The back-pressure turbine will operate with constant steam flow rate, but the steam flow rate to the ORC system varies with the steam/heat demand. The ORCs have the distinctive features of being simple cycle layouts, being low cost and especially being efficient for partial loads. Therefore, an ORC can generate power using the excess steam with high thermal efficiency and availability for complex operating conditions. In this conceptual system, the back-pressure steam turbine is operating efficiently at the optimal condition, and the extra electricity generated from the ORC system is sold to the grid which improves the economics considerably.

A typical B25–8.83/1.5 extraction back-pressure turbine is taken as a case study. The parameters for typical conditions of the back-pressure turbine are listed in **Table 7**, and the schematic is shown in **Figure 9**. The superheated steam from the boiler enters the back-pressure steam turbine to generate power. One stage steam is extracted from the steam turbine for the feedwater heater 1 (FWH1) to preheat the feedwater. In the original system, the exhaust steam of the back-pressure turbine is divided into four stages of steam flow: one stage extracted from the exhaust steam for FWH2 to preheat the feedwater, one stage extracted for the deaerator after throttling, one stage extracted for the mixer after throttling to a lower pressure to

Parameters	THA	75%THA	50%THA	30%THA
Main-steam temperature/°C	535	535	535	535
Main-steam pressure/MPa	8.83	8.83	8.83	8.83
Main-steam flow rate/$t{\cdot}h^{-1}$	245	195	148	109
Power output/kW	25,203	18,862	12,637	7545
Exhaust steam temperature/°C	316.2	327	345.3	370.6
Exhaust steam pressure/MPa	1.5	1.5	1.5	1.5
Steam demand/$t{\cdot}h^{-1}$	174.5	140.9	108.6	81.0
Isentropic efficiency/%	79.5	74.77	66.77	55.87

Table 7.
Operating parameters for a B25–8.83/1.5 back-pressure turbine [34].

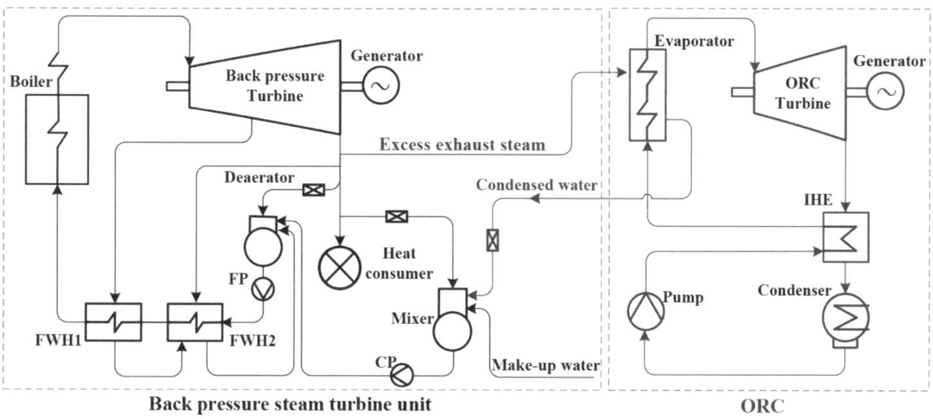

Figure 9.
Schematic of an ORC for recovering the excess exhaust steam heat from a back-pressure steam turbine [34].

preheat the makeup water, and the rest most of exhaust steam transported to the heat consumer. The heat consumer steam demand accounts for 70% of the turbine main-steam flow. The main-steam flow rate of the back-pressure turbine decreases with decreasing steam demand, which results in a serious decrease in the isentropic efficiency (relative internal efficiency) as shown in **Table 7**.

A conceptual system had been proposed to improve the thermodynamic performance of the back-pressure turbine as shown in **Figure 9** [34]. The back-pressure turbine is always operating at the turbine heat acceptance (THA) condition. When the heat consumer steam demand decreases, the excess exhaust steam is provided to an ORC for organic working fluid heating. The steam is condensed to the saturated water and then returns back to the mixer after throttling to preheat makeup water. Compared to the conditions with lower main steam flow rates, the back-pressure turbine generates additional electricity, $\Delta\dot{W}_{\mathrm{T}}$, due to the high isentropic efficiency and a constant steam flow rate. The amount of electricity produced by the ORC system using the excess exhaust steam is defined as $\dot{W}_{\mathrm{O}}$. Thus, the total additional electricity (increase in electricity) from the system can be expressed as

$$\Delta\dot{W}_{\mathrm{Sys}} = \Delta\dot{W}_{\mathrm{T}} + \dot{W}_{\mathrm{O}} \tag{10}$$

The additional electricity can be provided to the facilities or the grid. The steam/heat and electricity demands can then both be regulated by adjusting the main-steam flow rate. The additional electricity from the back-pressure turbine, $\Delta\dot{W}_{\mathrm{T}}$, mainly depends on the turbine operating parameters and the steam demand. The electricity produced from the ORC system, $\dot{W}_{\mathrm{O}}$, depends on the heat input from the excess exhaust steam and the ORC thermal efficiency [34].

The flow rates of the excess exhaust steam for various steam demands are determined according to the mass and the energy balances in the back-pressure turbine system. The steam in the ORC evaporator is condensed to the saturated liquid. The pressure drop between the back-pressure turbine outlet and the ORC evaporator outlet is set to be 5%. The superheat degree of the exhaust steam reaches 120°C, but the sensible heat is much lower than the latent heat. Thus, the matching characteristics between the organic working fluid evaporation and the steam condensation temperatures affect the ORC thermal efficiency.

The ORC turbine inlet temperature was assumed to be 195°C. The ORC turbine inlet pressures of the selected working fluids were optimized to maximize the net cycle power output with the boundary conditions listed in **Table 1**. The parameters and thermal efficiencies of ORCs are listed in **Table 8**. **Figure 10(a)** shows the *T-Q* diagrams of the evaporation process for the selected working fluids. Supercritical cycles are adopted for R134a, R1234ze, R236ea, and R245fa due to their lower critical temperatures and higher heat source temperature, while subcritical ORCs

Working fluid	Cycle	$T_{\mathrm{T,in}}$/°C	$p_{\mathrm{T,in}}$/MPa	$p_{\mathrm{T,out}}$/MPa	Thermal efficiency/%
R134a	s2	195	7.15	0.77	18.99
R1234ze	s2	195	6.09	0.58	19.40
R236ea	s2	195	4.37	0.24	20.59
R245fa	s2	195	3.87	0.18	20.93
R1233zd	o3	195	3.23	0.15	20.93
R123	o3	195	2.80	0.11	21.48

Table 8.
Comparison of ORCs using exhaust steam from a back-pressure turbine.

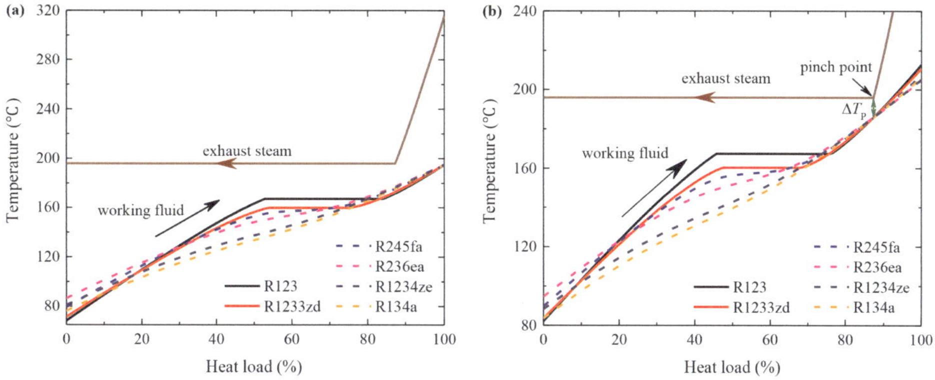

Figure 10.
T-Q diagrams for working fluid evaporating at (a) the optimal pressure and (b) both the optimized pressure and temperature.

with R123 and R1233zd are still used here. Subcritical ORC using R123 with the lowest evaporation pressure shows a better matching of the steam and the working fluid temperature profiles as shown in **Figure 10(a)** and the highest thermal efficiency among the working fluids. The working fluid with a lower critical temperature has a higher evaporation pressure but shows a lower thermal efficiency. The thermal efficiencies for R1233zd and R245fa are very close.

In this conceptual system, the back-pressure turbine outputs 6.3, 12.6, and 17.7 MWh additional electricity for the steam demands of 140.9, 108.6, and 81 t/h, respectively. The additional electricity generation for ORCs using the excess exhaust steam is listed in **Table 9**. As the steam demand decreases to 140.9 t/h, the ORCs generate about 6 MWh electricity with the excess steam flow rate of 44.8 t/h. The ORCs can generate 16.7 MWh electricity when the steam demand is decreased to 81 t/h. Thus, the total additional electricity output of the system, $\Delta \dot{W}_{Sys}$, can reach 12–34 MWh for partial steam demands.

At last, this section gives a further theoretical discussion on the cycle and working fluid choices for the heat source type with isothermal condensation. Both the temperature and the pressure at the ORC turbine inlet were optimized simultaneously to obtain the maximum power output. **Figure 10(b)** shows the T-Q diagrams for the working fluids evaporating at the optimal condition. The exhaust steam from the back-pressure turbine has a much large superheat degree, but the sensible heat is still much lower than the latent heat. The pinch point occurs at the dew point (saturated vapor state) of the steam as shown in **Figure 10(b)**. The working fluid may be heated to a temperature higher than the steam condensation temperature by the sensible heat. In this case study, the optimal temperatures of the working fluids at the evaporator outlet range from 205–213°C. The working fluid with a higher critical temperature has a higher optimal evaporator outlet temperature and then results in a higher thermal efficiency. The subcritical ORC using the

Steam demand/ $t \cdot h^{-1}$	$\dot{W}_O$/kW·h					
	R134a	R1234ze	R236ea	R245fa	R1233zd	R123
140.9	5292	5409	5741	5835	5835	5986
108.6	10,536	10,769	11,429	11,617	11,617	11,918
81.0	14,744	15,069	15,992	16,256	16,256	16,677

Table 9.
Electricity generated by ORCs using excess exhaust steam at various steam demands.

working fluid with a higher critical temperature is more suitable for the kind of isothermal heat source due to the better matches of the temperature profiles of the working fluid and the heat source fluid. The temperature at the evaporator outlet (turbine inlet) strongly affects the cycle thermal efficiency. The cycle efficiency can be relatively increased by about 4% for a 10°C increase in the turbine inlet temperature. Compared to the ORCs with the fixed turbine inlet temperature of 195°C, the cycle efficiencies are relatively improved by 3.7–5.7% for the optimal turbine inlet temperatures and pressures as shown in **Figure 11**. Among the six working fluids, the cycle efficiency with R123 is the highest with the highest turbine inlet temperature, followed by R1233zd. The ORC thermal efficiency can exceed 20% for a steam heat source at a temperature of 200°C. However, the thermal stabilities of the working fluids should be considered primarily [16, 18, 35].

5.2 ORCs driven by exhaust steam form a condensing steam turbine

The majority of heat loss in a steam Rankine cycle is the exhaust steam latent heat released to the coolant at the condenser. The cold source loss generally accounts for 50–60% of the total heat input for an extraction condensing turbine. The steam condensation temperature should be close to the coolant (environment) temperature to reduce the turbine exhaust pressure and then increase the thermal efficiency (absolute internal efficiency). The exhaust pressures generally range 4.9–11.8 kPa for condensing turbines using closed cooling systems with circulating water as the coolant; thus, the exhaust steam is condensed at temperatures of 32.5–49°C with the circulating water flow rate up to 50–100 times the exhaust steam flow rate. The condensation temperature (turbine exhaust pressure) is mainly affected by the environment temperature and the operating conditions. The seawater is generally used as the coolant for offshore power plants. The seawater temperature can reach as low as 5 °C in the cold season or the seawater from a depth up to 1000 m; thus, the turbine exhaust pressures range 3.5–5 kPa with the condensation temperatures between 26.7 and 32.9°C. Considering the issues including wetness fraction, volume flow rate, and blade length of the last stage of the low-pressure turbine (LPT), the exhaust pressure generally does not need to be less than 3.5 kPa even through the cooling water temperature is much lower. The ratio of the

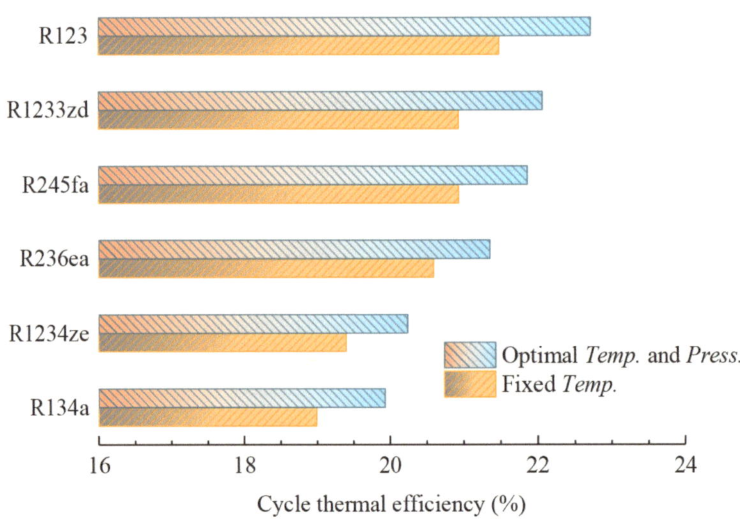

Figure 11.
Cycle thermal efficiencies for the ORCs using exhaust steam from back-pressure turbine.

circulating seawater flow rate to the exhaust steam flow rate can be reduced to 30–50 because of the higher cooling water temperature rise.

Organic Rankine cycles have been studied to utilize the ocean thermal energy with the temperature difference between warm seawater and cold seawater as low as 20°C [36–39]. Therefore, the temperature difference between the turbine exhaust steam and the cooling water can be used to drive an ORC for power generation as shown in **Figure 12**. The exhaust steam from the low-pressure turbine (LPT) is condensed by the organic working fluid in the condenser. The working fluid is heated to a saturated vapor and then enters into the ORC turbine to expand. After expansion, the organic working fluid is condensed by the cold seawater in the ORC condenser. The cycle thermal efficiency of the ORC is lower due to the lower temperature difference between the heat source and sink. However, the ORC can also provide substantial power because the discharged heat from the exhaust steam in a power plant is much huge.

The parameters and boundary conditions for the ORC using exhaust steam as the heat source and cold seawater as the heat sink are listed in **Table 10**. A subcritical 600 MW condensing steam turbine is taken as a case study. The exhaust pressure of the steam turbine is 3.5 kPa with a flow rate of 300 kg/s. The subcritical ORC with saturated vapor at the turbine inlet is only studied here due to the exhaust steam isothermal condensation. The organic working fluid evaporation temperature is only related to the steam condensation temperature with the pinch point temperature difference, ΔT_P; thus, the organic working fluid condensation temperature is the key parameter which affects the thermodynamic performance as shown in **Figure 13**. The ORC thermal efficiency increases as the condensation temperature decreases. A lower condensation temperature needs a lower cooling water (seawater) temperature rise with a higher flow rate; however, the cooling water pumps consume a significant fraction of the ORC output power [27, 36–39]. The cooling water (seawater) flow rate was assumed here to be 45,000 kg/s with a temperature rise about 3.58°C.

Figure 14 shows the turbine power generation and the net power output of the ORC system using the exhaust steam as heat source and the cold seawater as heat sink. The ORC turbine output can exceed 23.5 MW in this case study. The turbine

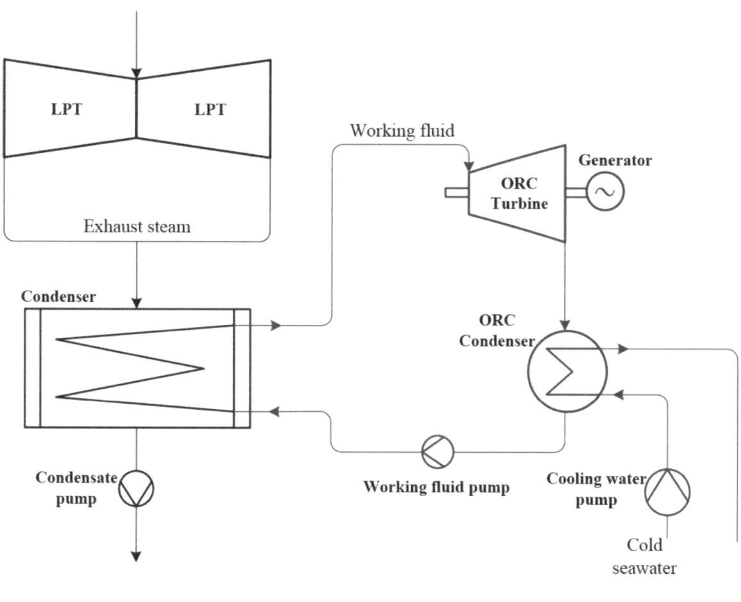

Figure 12.
Schematic of an ORC using the exhaust steam from a condensing turbine and the cold seawater.

Parameters	Values	Unit
Turbine isentropic efficiency [36]	80	%
Working fluid pump efficiency [36]	75	%
Cooling water pump efficiency	80	%
Generator efficiency	98	%
Turbine mechanical efficiency	98	%
Evaporator pinch point temperature difference [37]	2	°C
Condenser pinch point temperature difference	2	°C
Exhaust steam temperature	26.67	°C
Exhaust steam dryness fraction	92	%
Exhaust steam flow rate	300	kg/s
Cold seawater temperature	5	°C
Cooling water (seawater) flow rate	45,000	kg/s
Cooling water pump head	10	m
Seawater specific heat capacity [37]	4.025	kJ/(kg °C)

Table 10.
Operating parameters and boundary conditions for the ORC using exhaust steam as heat source and seawater as sink.

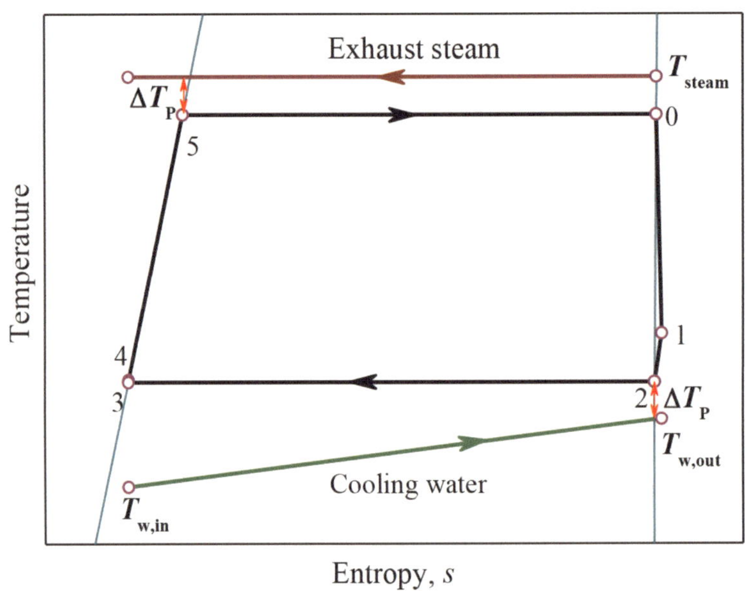

Figure 13.
Temperature-entropy diagram of an ORC using R1234ze as the working fluid driven by exhaust steam from a condensing turbine with cold seawater as heat sink.

using R134a produces the maximum power among the selected six working fluids. The cooling water pumps consume a significant fraction of the turbine output power. The pump power consumption accounts for about 23.3% of the turbine power generation when the cooling water pump head is 10 m. The pressure increase in the working fluid pump is much higher for the working fluid with a lower critical temperature and leads to a higher power consumption. The parasitic power consumed by the working fluid feed pump with R134a accounts for 3.6% of the turbine

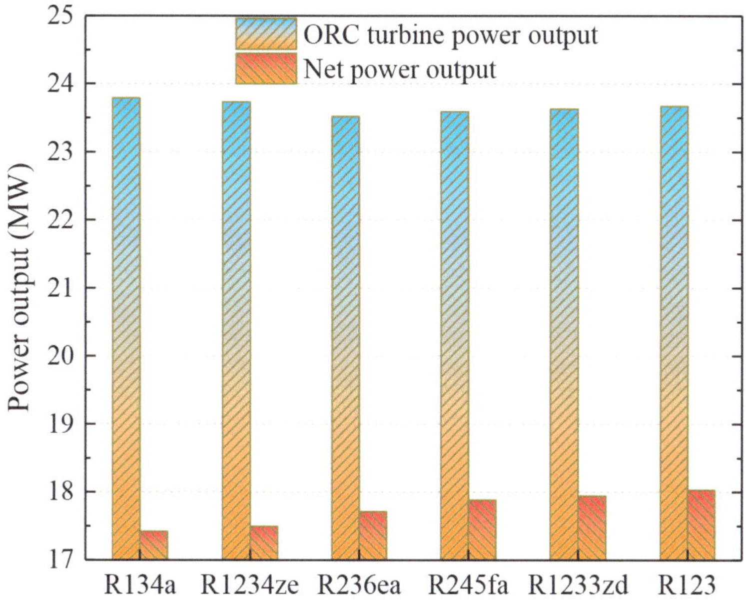

Figure 14.
Turbine power output and net cycle power output for ORCs with exhaust steam as heat source and cold seawater as heat sink.

power output. Thus, the system net power output (subtracting the power consumed by the working fluid pump and the cooling water pumps from the turbine power output) with R134a is the lowest due to the largest parasitic power consumption, while the ORC with R123 outputs the maximum net power among the six working fluids. When the cold seawater is directly used to condense the exhaust steam from LPTs, the cooling water flow rate for the steam turbine is about 8,500 kg/s, but the power consumed by the circulating water pumps is still more than 1 MW. When the exhaust steam is used as an ORC heat source with the cold seawater as the heat sink, the system can provide net power output up to 18 MW equal to 3% increase in the gross power of the coal-fired power unit.

6. Conclusions

In coal-fired power plants, there are a large number of medium-low-grade waste heat which is originally discharge to the environment. Conceptual systems and thermodynamic performance for the ORCs recovering waste heat from coal-fired power plants are studied in this chapter. The optimal cycle layouts and proper working fluids for the ORCs are discussed. The conclusions are summarized as follow.

1. The amount of the drained water that directly discharges has been toward fewer and smaller in modern power plants. The original discharge or leaked drain can be collected to drive a small-capacity ORC. A supercritical ORC matches well with the kind of heat source with a large temperature drop.

2. The flue gas flow rate is much high for a large-capacity boiler. The amount of waste heat in exhaust flue gas can be reached 10 MW even through the flue gas is only cooled by 10°C. The subcritical ORC with an IHE (a recuperator or regenerator) using a working fluid with a higher critical temperature is more

suitable for the kind of heat source with a higher limited outlet temperature. The gross power output of the coal-fired power unit can be relatively increased by 0.4% by the ORC using the waste heat from boiler exhaust flue gas.

3. More than 50% of the heat input to the turbine is released in the condenser. When the cooling water temperature is much lower, the use of ORCs to recover the waste heat from the turbine exhaust steam becomes attractive. Although the cooling water pumps consume a significant fraction of the ORC output power, the net efficiencies of ORCs still reach 2.6%. The power output increase potential for the coal-fired power unit can exceed 2–3% when all of the turbine exhaust steam is used to drive ORCs with lower temperature seawater as the heat sink.

Acknowledgements

This work was supported by the National Natural Science Foundation of China (Grant Nos. 51506223 and 51736005) and the National Basic Research Program of China (973 program) (Grant No. 2015CB251502).

Conflict of interest

The authors declare no competing financial interest.

Author details

Qiang Liu
College of Mechanical and Transportation Engineering, China University of Petroleum, Beijing, P.R. China

*Address all correspondence to: qliu@cup.edu.cn

References

[1] Available from: https://www.iea.org/geco/data/ [Accessed: 08 June 2019]

[2] Available from: https://github.com/citation-style-language/styles/blob/master/vancouver-brackets.csl [Accessed: 08 June 2019]

[3] Che DF. Boilers - Theory, Design and Operation. Xi'an: Xi'an Jiaotong University Press; 2008

[4] Liu Q, Duan YY, Wan XC. Power generation system using continuous blowdown waste heat from drum boiler driving an organic Rankine cycle. Proceedings of the CSEE. 2013;**33**(35): 1-7. DOI: 10.13334/j.0258-8013.pcsee.2013.35.006

[5] Yang YP, Xu C, Xu G, Han Y, Fang YX, Zhang DK. A new conceptual cold-end design of boilers for coal-fired power plants with waste heat recovery. Energy Conversion and Management. 2015;**89**:137-146. DOI: 10.1016/j.enconman.2014.09.065

[6] Liu M, Zhang XW, Ma YG, Yan JJ. Thermo-economic analyses on a new conceptual system of waste heat recovery integrated with an S-CO2 cycle for coal-fired power plants. Energy Conversion and Management. 2018;**161**: 243-253. DOI: 10.1016/j.enconman.2018.01.049

[7] Wang CJ, He BS, Yan LB, Pei XH, Chen SN. Thermodynamic analysis of a low-pressure economizer based waste heat recovery system for a coal-fired power plant. Energy. 2014;**65**:80-90. DOI: 10.1016/j.energy.2013.11.084

[8] Łukowicz H, Kochaniewicz A. Analysis of the use of waste heat obtained from coal-fired units in organic Rankine cycles and for brown coal drying. Energy. 2012;**45**:203-212. DOI: 10.1016/j.energy.2012.03.035

[9] Liu M, Zhang XW, Han XQ, Li G, Yan JJ. Using pre-drying technology to improve the exergetic efficiency of bioenergy utilization process with combustion: A case study of a power plant. Applied Thermal Engineering. 2017;**127**:1416-1426. DOI: 10.1016/j.applthermaleng.2017.08.156

[10] Han XQ, Liu M, Wu KL, Chen WX, Xiao F, Yan JJ. Exergy analysis of the flue gas pre-dried lignite-fired power system based on the boiler with open pulverizing system. Energy. 2016;**106**:285-300. DOI: 10.1016/j.energy.2016.03.047

[11] Han XQ, Karellas S, Liu M, Braimakis K, Chen WX, Yan JJ, et al. Integration of organic Rankine cycle with lignite flue gas pre-drying for waste heat and water recovery from dryer exhaust gas: Thermodynamic and economic analysis. Energy Procedia. 2017;**105**:1614-1621. DOI: 10.1016/j.egypro.2017.03.518

[12] Huang SW, Li CZ, Tan TY, Fu P, Wang LG, Yang YP. Comparative evaluation of integrated waste heat utilization systems for coal-fired power plants based on in-depth boiler-turbine integration and organic Rankine cycle. Entropy. 2018;**20**:89. DOI: 10.3390/e20020089

[13] Xi XM, Zhou YY, Guo C, Yang LJ, Du XZ. Characteristics of organic Rankine cycles with zeotropic mixture for heat recovery of exhaust gas of boiler. Energy Procedia. 2015;**75**: 1093-1101. DOI: 10.1016/j.egypro.2015.07.496

[14] Mikielewicz D, Wajs J, Ziółkowski P, Mikielewicz J. Utilisation of waste heat from the power plant by use of the ORC aided with bleed steam and extra source of heat. Energy. 2016; **97**:11-19. DOI: 10.1016/j.energy.2015.12.106

[15] Angelino G, Invernizzi C, Molteni G. The potential role of organic bottoming Rankine cycles in steam power stations. Proceedings of the Institution of Mechanical Engineers, Part A: Journal of Power and Energy. 1999;**213**:75-81. DOI: 10.1243/0957650991537446

[16] Macchi E. Theoretical basis of the organic Rankine cycle. In: Macchi E, Astolfi M, editors. Organic Rankine Cycle (ORC) Power Systems. London: Woodhead Publishing; 2017. pp. 3-24. DOI: 10.1016/B978-0-08-100510-1.00001-6

[17] Saleh B, Koglbauer G, Wendland M, Fischer J. Working fluids for low-temperature organic Rankine cycles. Energy. 2007;**32**:1210-1221. DOI: 10.1016/j.energy.2006.07.001

[18] Invernizzi CM. Closed Power Cycles - Thermodynamic Fundamentals and Applications. London: Springer-Verlag; 2013. DOI: 10.1007/978-1-4471-5140-1

[19] Astolfi M. Technical options for organic Rankine cycle systems. In: Macchi E, Astolfi M, editors. Organic Rankine Cycle (ORC) Power Systems. London: Woodhead Publishing; 2017. pp. 67-89. DOI: 10.1016/B978-0-08-100510-1.00003-X

[20] Madhawa Hettiarachchi HD, Golubovic M, Worek WM, Ikegami Y. Optimum design criteria for an organic Rankine cycle using low-temperature geothermal heat sources. Energy. 2007; **32**:1698-1706. DOI: 10.1016/j.energy.2007.01.005

[21] Maraver D, Royo J, Lemort V, Quoilin S. Systematic optimization of subcritical and transcritical organic Rankine cycles (ORCs) constrained by technical parameters in multiple applications. Applied Energy. 2014;**117**: 11-29. DOI: 10.1016/j.apenergy.2013.11.076

[22] Quoilin S, Broek MVD, Declaye S, Dewallef P, Lemort V. Techno-economic survey of organic Rankine cycle (ORC) systems. Renewable and Sustainable Energy Review. 2013;**22**: 168-186. DOI: 10.1016/j.rser.2013.01.028

[23] Zhai HX, An QS, Shi L, Lemort V, Quoilin S. Categorization and analysis of heat sources for organic Rankine cycle systems. Renewable and Sustainable Energy Review. 2016;**64**:790-805. DOI: 10.1016/j.rser.2016.06.076

[24] Sun J, Liu Q, Duan YY. Effects of evaporator pinch point temperature difference on thermo-economic performance of geothermal organic Rankine cycle systems. Geothermics. 2018;**75**:249-258. DOI: 10.1016/j.geothermics.2018.06.001

[25] Pan LS, Shi WX. Investigation on the pinch point position in heat exchangers. Journal of Thermal Science. 2016;**25**:258-265. DOI: 10.1007/s11630-016-0858-7

[26] Liu Q, Duan YY, Yang Z. Performance analyses of geothermal organic Rankine cycles with selected hydrocarbon working fluids. Energy. 2013;**63**:123-132. DOI: 10.1016/j.energy.2013.10.035

[27] Liu Q, Shen AJ, Duan YY. Parametric optimization and performance analyses of geothermal organic Rankine cycles using R600a/R601a mixtures as working fluids. Applied Energy. 2015;**148**:410-420. DOI: 10.1016/j.apenergy.2015.03.093

[28] Xu JL, Yu C. Critical temperature criterion for selection of working fluids for subcritical pressure organic Rankine cycles. Energy. 2014;**74**:719-733. DOI: 10.1016/j.energy.2014.07.038

[29] Lemmon EW, Huber ML, McLinden MO. NIST Standard Reference Database 23: Reference Fluid

Thermodynamic and Transport Properties-REFPROP, Version 9.1. Gaithersburg: National Institute of Standards and Technology, Standard Reference Data Program; 2013

[30] Rayaprolu K. Boilers for Power and Process. Boca Raton: Taylor & Francis Group; 2009

[31] Liu Q, Shang LL, Duan YY. Performance analyses of a hybrid geothermal–fossil power generation system using low-enthalpy geothermal resources. Applied Energy. 2016;**162**: 149-162. DOI: 10.1016/j.apenergy.2015. 10.078

[32] Breeze P. Combined Heat and Power. London: Elsevier; 2018. DOI: B978-0-12-812908-1.00003-1

[33] Ohji A, Haraguchi M. Steam turbine cycles and cycle design optimization: The Rankine cycle, thermal power cycles, and IGCC power plants. In: Tanuma T, editor. Advances in Steam Turbines for Modern Power Plants. London: Woodhead Publishing; 2017. pp. 11-40. DOI: 10.1016/B978-0-08-100314-5.00002-6

[34] Liu Q, Duan YY. Cogeneration system coupled with a back pressure steam turbine unit and an organic Rankine cycle (ORC). Proceedings of the CSEE. 2013;**33**(23):29-35. DOI: 10.13334/j.0258-8013.pcsee.2013.23.014

[35] Dai XY, Shi L, Qian WZ. Review of the working fluid thermal stability for organic Rankine cycles. Journal of Thermal Science. 2019;**28**:597-607. DOI: 10.1007/s11630-019-1119-3

[36] Yamada N, Hoshi A, Ikegami Y. Performance simulation of solar-boosted ocean thermal energy conversion plant. Renewable Energy. 2009;**34**:1752-1758. DOI: 10.1016/j. renene.2008.12.028

[37] Aydin H, Lee HS, Kim HJ, Shin SK, Park K. Off-design performance analysis of a closed-cycle ocean thermal energy conversion system with solar thermal preheating and superheating. Renewable Energy. 2014;**72**:154-163. DOI: 10.1016/j.renene.2014.07.001

[38] Khan N, Kalair A, Abas N, Haider A. Review of ocean tidal, wave and thermal energy technologies. Renewable and Sustainable Energy Review. 2017;**72**: 590-604. DOI: 10.1016/j.rser.2017. 01.079

[39] Bernardoni C, Binotti M, Giostri A. Techno-economic analysis of closed OTEC cycles for power generation. Renewable Energy. 2019;**132**:1018-1033. DOI: 10.1016/j.renene.2018.08.007

Chapter 4

The Development and Application of a Small-Scale Organic Rankine Cycle for Waste Heat Recovery

Tzu-Chen Hung and Yong-qiang Feng

Abstract

Power conversion systems based on organic Rankine cycles have been identified as a potential technology especially in converting low-grade waste heat into electricity as well as in small-scale biomass, solar, or geothermal power plants. The theoretical analysis can guide the ORC design, but cannot predict accurately the system performance. Actually, the operation characteristics of every component have a vital effect on the system performance. This chapter presents the detailed operation characteristic of a small-scale ORC. The effects of the operation parameters, the mixture working fluid and the operation strategy on system overall performance are addressed. It can be concluded that improving the system overall performance should give priority to increase the pressure drop. Whether the mixtures exhibit better thermodynamic performance than the pure working fluids depend on the operation parameters and mass fraction of mixtures. The mixture working fluids obtain a higher expander shaft power but a relatively higher BWR. The expander rotating speed for standalone operation strategy keeps rising from 2320 to 2983 rpm, whereas that of grid connect operation strategy keeps constant of 3600 rpm.

Keywords: organic Rankine cycle (ORC), operation characteristic, mixture working fluids, system generating efficiency

1. Introduction

Energy is an indispensable resource for human progress and social development, improving energy efficiency has become a global research hot spot. Meanwhile, waste heat resource utilization problem has received widespread attention. If those waste heats can be effectively utilized, it will not only provide important technical support for energy conservation, emission reduction and environmental protection, but also generate certain economic benefits. Organic Rankine cycle (ORC) was adapted as a new technology to utilize waste heat [1–3]. The principle of ORC is similar to that of Rankine cycle. The main difference is that the low-boiling organic fluid can be used to replace the water of the Rankine cycle, which can significantly reduce the final discharge temperature, thus achieving the purpose of waste heat recovery [4–8].

Recently, many scholars have conducted in-depth research on ORC system concerning on the working fluids selection, ORC performance optimization and component development. However, the theoretical analysis can guide the ORC design, but cannot predict accurately the system performance. Therefore, some researchers devoted main effort on ORC experimental studies.

Mathias et al. [9] compared the operation characteristics using piston pump and gear pump, expressing that the piston pump outperformed than the gear pump. Carraro et al. [10] integrated a multi-diaphragm positive displacement pump into a 4 kW ORC experimental prototype, and found that the pump global efficiency was about 45–48%. Xu et al. [11] used R123 to study the matching degree of the piston pump and the expander. They stated that the low pump frequency was applicable to all expander torques, while the high pump frequency was only applicable to low torques. Zhao et al. [12] studied the diaphragm pump performance under various conditions using four different working fluids. The experimental results showed that a higher volume flow rate and pressure difference led to a higher pump isentropic efficiency. Lei et al. [13] tested a roto-jet pump at different rotating speeds using R123. They illustrated that the pump efficiency was in range of 11–23%, and an increase in pump rotating speed or a decrease in mass flow rate can cause the decrease in pump efficiency. Zeleny et al. [14] proposed a modified gear pump used in micro ORC and discussed the contribution to pump different losses. They stated that as the pressure increases, the effect of mechanical losses decreased, while the volume loss was reversed. Bianchi et al. [15] applied a sliding vane pump on the ORC using R236fa. When the mass flow rate increased from 52 to 119 g/s, the pump's mechanical power increased by 289 W. Wu et al. [16] used multistage gas-liquid booster pump to improve the performance of ORC system, revealing that the maximum conversion efficiency from high-pressure air to water was 0.72. Yang et al. [17] compared experimental characteristics of three pumps, demonstrating that the maximum actual pump efficiency of multistage centrifugal pump was 58.76%. Landelle et al. [18] discussed the operation characteristic of the reciprocating pump, reporting that the reciprocating pump had eminent volumetric efficiency and the ORC efficiency decreased as cavitation margin increased. Meng et al. [19] analyzed the multistage centrifugal pump on the ORC, expressing that the maximum overall pump efficiency was 65.7%. Bianchi et al. [20] changed the heat source temperature and feed pump speed to explore the performance of micro ORC. They found that the achieved pump efficiency was ranging from 10 to 20% and the net efficiency of the system was 2.2%. Yang et al. [21] studied the key parameters of the hydraulic diaphragm metering pump. The tested efficiency and BWR were 88.7% and 0.93, respectively. Bianchi et al. [22] used the CFD model to design and analyze a sliding vane pump. The pump performance was found to be optimal at 1250 rpm and 9.7 bar. Sun et al. [23] compared three different pumps from the viewpoint of the actual cycle and the ideal cycle. The experimental results showed that the combination of a centrifugal pump and a scroll expander can maximize the isentropic efficiency of the expander. Xi et al. [24] analyzed the influence of the plunger stroke of the working fluid pump on the system performance by orthogonal analysis. The results showed that the plunger stroke had a great influence on pump power consumption and working fluid flow. Aleksandra Borsukiewicz-Gozdur [25] found that a higher pumping power required a higher cycle pressure.

Compared with the pure working fluids, the main advantage of mixtures as ORC working fluids stems from their non-isothermal phase transitions during vaporization and condensation, and hence effectively match the temperature change of heat source and cooling water. Therefore, great attention has been drawn to the mixture working fluids. Dong et al. [26] compared the thermal efficiency of high-temperature ORC system between mixture and pure working fluids, and found that the range of options for working fluids was widened by the mixture working fluids. Garg et al. [27] investigated the thermodynamic analysis using isopentane, R245fa and their mixtures, reporting that 0.7isopentane/0.3R245fa was the preferred candidate working fluid. Lecompte et al. [28] discussed the exergy efficiency of ORC system using mixture working fluids, stating that the mixture presented 7.1–14.2% higher second law efficiency than the pure working fluids. Zhao et al. [29] analyzed the effect of composition shift of mixture working fluids on system performance, demonstrating that the composition shift significantly influenced the performance of ORC system. Liu et al. [30] investigated the effect of condensation temperature glide using mixture on ORC performance, reporting that when the increase of cooling water temperature is greater than the condensation temperature glide, an optimal working fluid mole fraction can be obtained to maximum thermodynamic performance. Furthermore, great attention has been drawn to the experimental comparison between mixture and pure working fluids. Pu et al. [31] compared the system performance using R245fa and HFE7100, showing that R245fa obtained the maximum net power output of 1.98 kW, which is 0.95 kW higher than that of HEF7100. Molés et al. [32] proposed using HCFO-1233zd-E to replace HFC-245fa, and found that the net electrical efficiency was in rage of 5–9.7%. Jung et al. [33] studied the dynamic behavior of a kW ORC test rig using R245fa/365mfc. Li et al. [34] compared the system behaviors using R245fa and R245fa/R601a.

In this chapter, the effect of mass flow rate, pressure drop, degree of superheating and condenser temperature on thermal efficiency and system generating efficiency are examined [35–36]. Several experimental investigations using pure working fluids (R123 and R245fa) of small-scale ORC test rig have been performed. However, few of them tried to fulfill the experimental comparison between pure and mixture working fluids. And therefore, two pure working fluids (R245fa, R123) and two mixtures working fluids (0.67R245fa/0.33R123 and 0.33R245fa/0.67R123) are tested and compared [37]. The system behaviors at two different operation strategies are addressed [38].

2. Experimental setup of ORC system

A 3 kW ORC experimental prototype is adopted, as shown in **Figure 1**. It includes three loops: heating loop, ORC loop and cooling loop. The photo of experimental layout and the main facilities of ORC system are plotted in **Figure 2**.

2.1 Heating loop

An electric heater using conduction oil is used as the simulated heat source, within four electrical heating rods having the capacity of 80 kW. An axial pump adjusted the mass flow rates of conductive oil, ensuring the heat source temperatures ranging from 110 to 140°C. Meanwhile, the evaporator heat transfer rate can be changed by adjusting the input power of electric heater, which is controlled by the four electrical heating rods.

2.2 ORC loop

The ORC loop is made up of four major components: a plunger pump, an evaporator, a scroll-type expander and a condenser. It should be noted that R123 is used in this study because of its better thermal efficiency and environmental performance. The pump supplies the working fluid to the evaporator where the working fluid is heated and vaporized by the conductive oil. The high pressure vapor flows into the expander and its enthalpy is converted into work. The low pressure vapor exits the expander and is led to the condenser where it is liquefied by water. The liquid is available at the condenser outlet, and then it is pumped back to the evaporator and a new cycle begins.

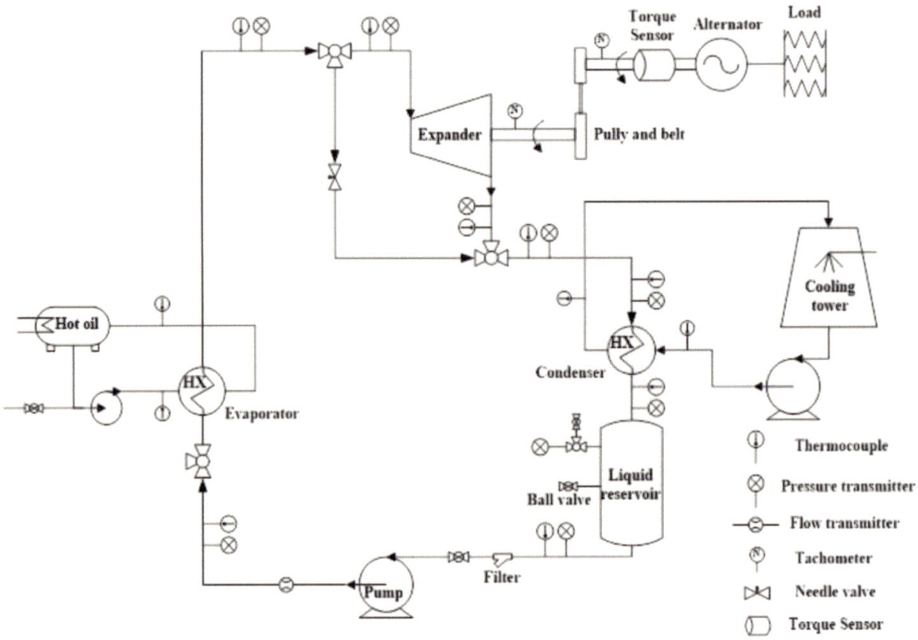

Figure 1.
Schematic diagram of an ORC system.

Figure 2.
The photos of (a) experimental setup, (b) the expander generator set, and (c) the electrical load.

The plunger pump is controlled by a frequency converter, with the achievable maximum delivered pressure head and flow rate are 50 bars and 18 L/min, respectively. The pump power consumption can be obtained by measuring the voltage and current, while the pump shaft power can be calculated by the thermodynamic parameters at the inlet and outlet of pump.

The evaporator and condenser are plate heat exchangers with the same heat transfer area of 4.157 m^2. Asbestos board and insulating foam are equipped around the evaporator and condenser to avoid the heat loss.

A scroll-type expander, which was modified from commercial oil-free scroll type air compressor with a built-in volume ratio of three, is employed. The expander shaft power is transferred to the three-phase permanent-magnet generator by pulley and belt. The alternating current produced by permanent-magnet generator is converted to direct current using three bridge rectifier. The frequency of direct current is adjusted by the electrical resistance and capacitance to meet the grid frequency requirements.

2.3 Cooling loop

The cooling tower is installed at roof and extras the heat from condenser to air environment. Therefore, the cooling inlet water is fluctuated by the environmental temperature. The needle value is used to adjust the mass flow rate of cooling water.

3. Thermodynamic analysis method

It should be noted that the thermodynamic properties of pure and mixture working fluids are obtained based on NIST Refprop [39]. Based on the measured temperatures and pressures, the corresponding enthalpies and entropies for every state can be obtained. **Figure 3** shows *T-s* plot of the thermodynamic cycle. For the expander, the vapor working fluid (state 1) enters the expander to generate power output, and then exits (state 2). The evaporator heat transfer rate (Q_{eva}), which is heated by the conductive oil, is

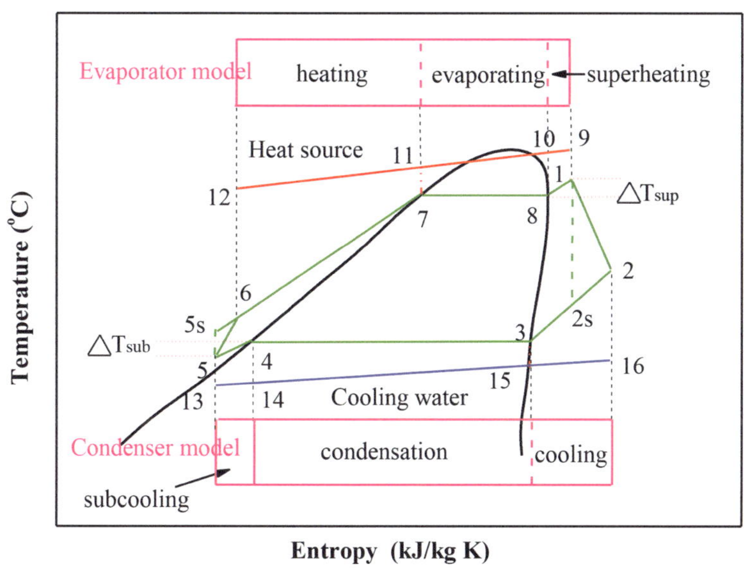

Figure 3.
T-s plot of the ORC cycle, as well as the evaporator and condenser model.

$$Q_{eva} = m(h_1 - h_6) \tag{1}$$

The expander power output is calculated according to the thermodynamic state at expander inlet and outlet, while the expander shaft power is measured by torque meter, which can be obtained by torque and rotating speed. Meanwhile, the ideal isentropic expansion process in expander is from 1 to 2 s, while the real one is from 1 to 2. The expander power output ($W_{p,exp}$) and shaft power ($W_{sh,exp}$) can be expressed as follows:

$$W_{p,exp} = m(h_1 - h_2) \tag{2}$$

$$W_{sh,exp} = \frac{2\pi}{60} M_{exp} n_{exp} \tag{3}$$

where h_1 and h_2 are the inlet and outlet enthalpy of expander, which is determined by the measured pressure and temperature; M_{exp} is the torque, and n_{exp} is the rotating speed, which is measured by tachometer.

The electrical power produced by generator can be calculated by measuring the current and voltage. The pump shaft power can be calculated by the measured pressure and temperature, while the power consumption is measured by pump frequency converter. To better understand the portion of the electricity output for driving the pump, back work ratio (BWR) is proposed to represent the ratio between pump consumption and electricity output, which can be expressed as:

$$\text{BWR} = \frac{W_{ele,pump}}{W_{ele,exp}} \tag{4}$$

The thermal efficiency and system generating efficiency can be expressed as:

$$\eta_{th,cal} = \frac{(h_1 - h_2) - (h_6 - h_5)}{h_1 - h_6} \tag{5}$$

$$\eta_{th,test} = \frac{W_{sh} - W_p}{Q_{eva}} \tag{6}$$

$$\eta_{ele} = \frac{W_{ele,exp} - W_{ele,p}}{Q_{eva}} \tag{7}$$

4. Effects of different operation parameters

The operation parameters have a significant influence on system performance. Based on the 3 kW ORC experimental prototype, the steady-state operation test is discussed at first, and then the four operation parameters on system behavior are examined.

4.1 Steady-state operation test

The heat input is used to control the heat source temperature, while the working fluid pump frequency is adopted to adjust the working fluid mass flow rate. The experimental data is collected at every 5 s. The test is recorded when the heat source temperatures are within a small fluctuation below ±0.5°C. One point is the average value of a 20 min steady operation. When the mass flow rate is in range

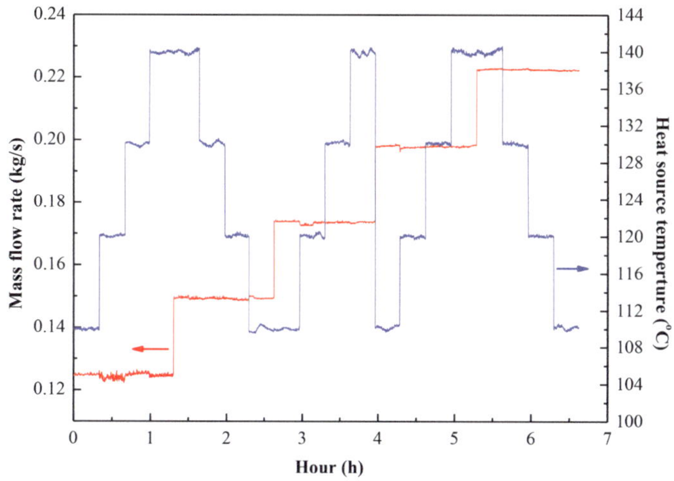

Figure 4.
Variation of heat source temperatures and mass flow rates with time.

of 0.124–0.222 kg/s and heat source temperature is in range of 110–140°C, the variation of heat source temperatures and mass flow rates with time are plotted in **Figure 4**. Obviously, the mass flow rate and heat source temperature have a relatively stable variation.

Taking the mass flow rate of 0.124 kg/s and heat source temperature of 110°C as an example, the detailed steady-state operation characteristic for 20 min steady operation is shown in **Figure 5**. It can be seen that the pump outlet pressure (**Figure 5b**), expander inlet pressure (**Figure 5c**), and mass flow rate of working fluids (**Figure 5f**) have a relativelŞy strong variation, indicating that a control strategy for the pump and expander is necessary.

4.2 Effect of mass flow rate

The working fluid mass flow rate is adjusted by the pump frequencies. **Figure 6** plots the variation of thermal efficiency with mass flow rate at different heat source temperatures. It can be seen that the thermal efficiencies for different heat source temperatures have a similar behavior of a decreasing trend with mass flow rate, which may be attributed to the increasing heat input. Meanwhile, the thermal efficiency keeps rising with the heat source temperature, which may be attributed to the increasing power output. It also can be found that when the mass flow rate is 0.124 kg/s and the heat source temperature is 140°C, the thermal efficiency owns a maximum value of 5.14%.

Figure 7 shows the variation of system generating efficiency with mass flow rate at different heat source temperatures. As the mass flow rate increases, the system generating efficiency keeps decreasing for a heat source temperature smaller than 120°C, whereas represents a parabolic trend for a heat source temperature higher than 130°C. Meanwhile, the system generating efficiency keeps rising with the heat source temperature. A highest system generating efficiency of 3.25% is appeared for a mass flow rate of 0.198 kg/s and a heat source temperature of 140°C. The system generating efficiency is in range of 0.94–3.25%.

4.3 Effect of pressure drop

The pressure drop denotes the expander inlet pressure minus the pump inlet pressure. The variation of electrical power, thermal efficiency and system generating

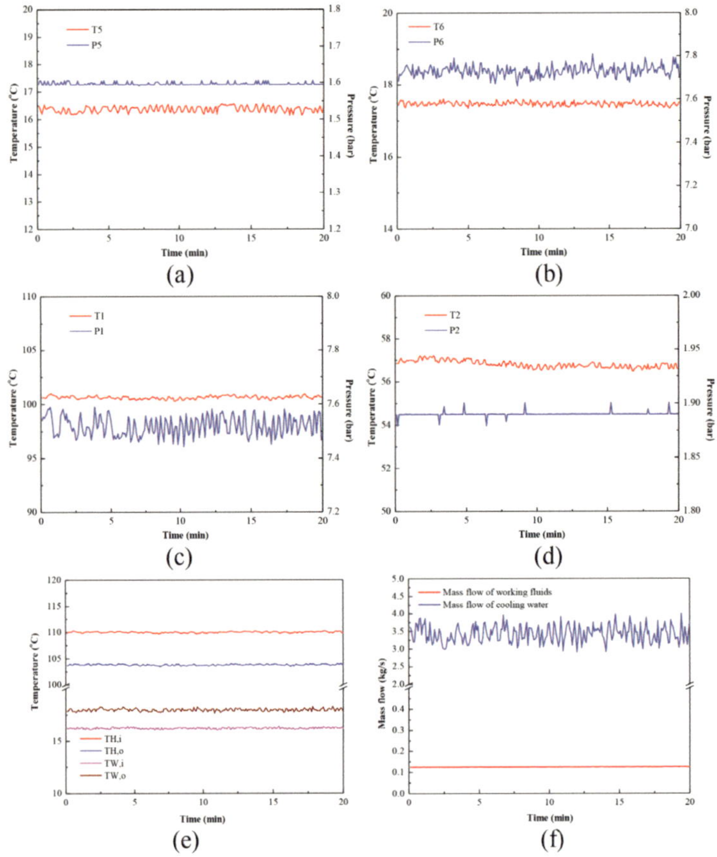

Figure 5.
Variation of system parameters with time. (a) pump inlet pressure and temperature, (b) pump outlet pressure and temperature, (c) expander inlet pressure and temperature, (d) expander outlet pressure and temperature, (e) cooling water inlet and outlet temperature and heat source inlet and outlet temperature, (f) mass flow rates of working fluid and cooling water.

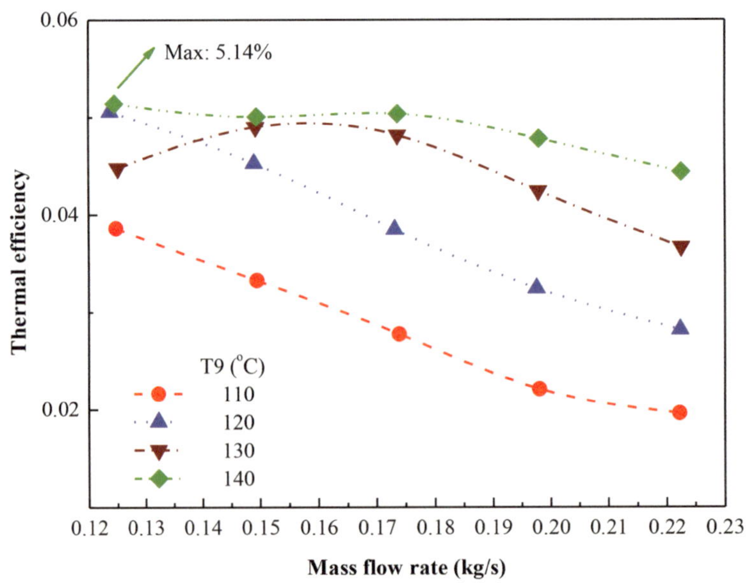

Figure 6.
Variation of thermal efficiency with mass flow rate at different heat source temperatures.

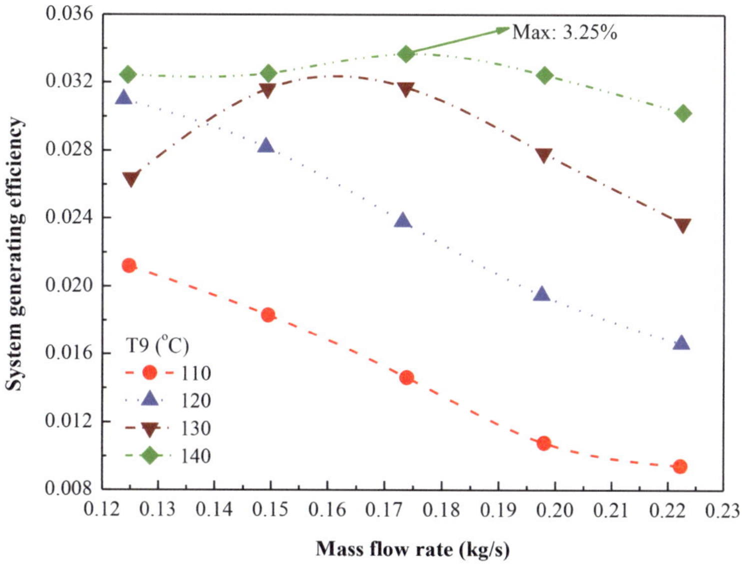

Figure 7.
Variation of system generating efficiency with mass flow rate at different heat source temperature.

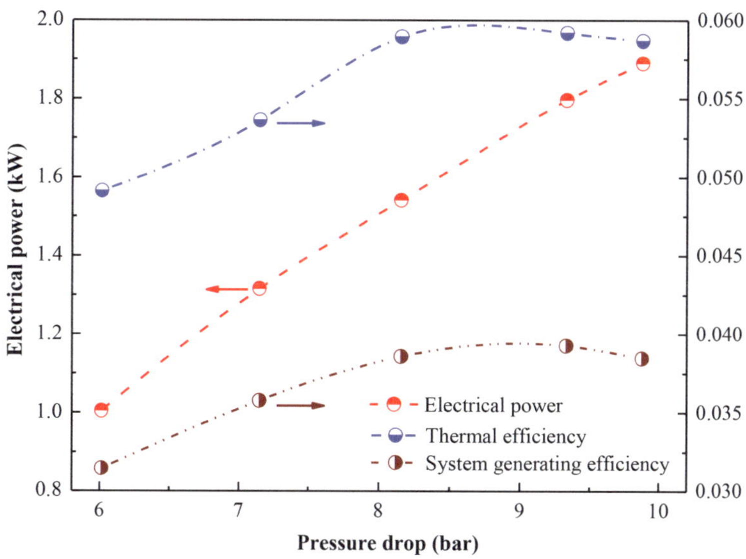

Figure 8.
Variation of electrical power, thermal efficiency and system generating efficiency with pressure drop.

efficiency are demonstrated in **Figure 8**. As the pressure drop increases, the system generating efficiency and thermal efficiency own a similar behavior of an increase first and then a slightly decrease, whereas the electrical power goes up. The increasing heat input causes the parabolic trend for thermal efficiency and system generating efficiency. A higher pressure drop denotes a higher investment cost of heat exchanger for pressure-bearing requirement. And therefore, the optimum pressure drop is 8.16 bar, with the corresponding thermal efficiency of 5.89% and system generating efficiency of 3.86%. It also can be found that the maximum thermal efficiency of 5.92% and maximum system generating efficiency of 3.93% are obtained with the corresponding pressure drop of 9.33 bar.

4.4 Effect of degree of superheating

The variation of electrical power, thermal efficiency and system generating efficiency with degree of superheating are displayed in **Figure 9**. It can be seen that the electrical power, thermal efficiency and system generating efficiency yield a small variation with the degree of superheating, which is in line with the theoretical study. The main reason is that the degree of superheating has a negligible effect on the power output. The electrical power of 1.35 kW, thermal efficiency of 6.48% and system generating efficiency of 3.91% can be owned.

4.5 Effect of condenser temperature

The needle value is used to adjust the cooling water mass flow rate, while the working fluid mass flow rate is set to be 0.10 kg/s. The variation of electrical power, thermal efficiency and system generating efficiency with condenser temperature are

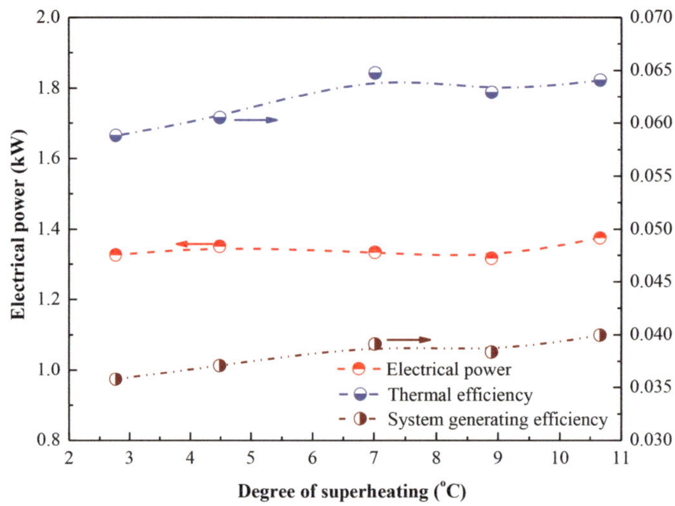

Figure 9.
Variation of electrical power, thermal efficiency and system generating efficiency with degree of superheating.

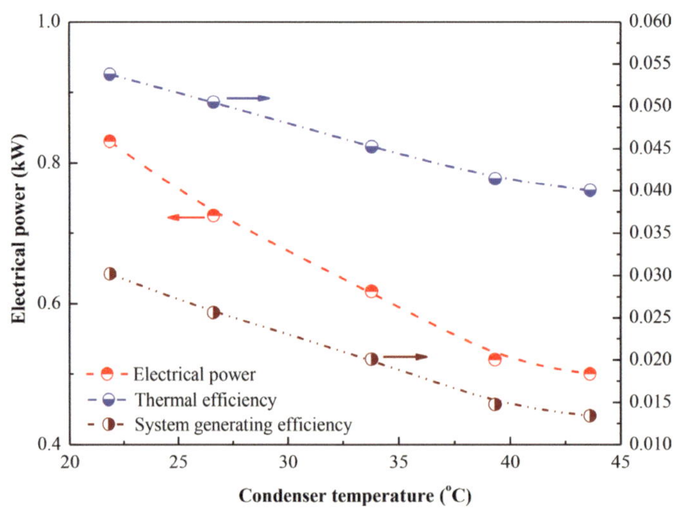

Figure 10.
Variation of electrical power, thermal efficiency and system generating efficiency with condenser temperature.

demonstrated in **Figure 10**. As the condenser temperature increases, the electrical power, thermal efficiency and system generating efficiency own a similar behavior of a decrease trend. The main reason is that the rising condenser temperature decreases the expander enthalpy difference, resulting in the decrease in the power output. When the condenser temperature increases from 22 to 43°C, the electrical power decreases from 0.83 to 0.50 kW, while the thermal efficiency decreases from 3.02 to 1.34%.

Based on the parametric analysis, the pressure drop has a relatively effect on the system performance, indicating that improving the system overall performance should give priority to increase the pressure drop. Meanwhile, the optimum electrical power and thermal efficiency are 1.89 kW and 5.92%, respectively. The maximum thermal efficiency does not represent the highest electrical power, which is in line with the theoretical study.

5. Operation characteristics comparison between mixture and pure working fluids

To better compare the operation characteristics between the pure working fluids and mixture working fluids, two pure working fluids (R245fa, R123) and two mixtures working fluids (0.67R245fa/0.33R123 and 0.33R245fa/0.67R123) are tested.

BWR denotes the ratio between pump consumption and electricity output. For theoretical study, the pump consumption is always ignore, whereas it accounts for a large proportion for ORC experimental prototype. The BWR for R245fa, R123 and their mixtures are displayed in **Figure 11**. As the mass flow rate increases, the BWR for different working fluids own a parabolic trend. There is an optimum mass flow rate to the lowest BWR. Meanwhile, the mixture working fluids yield a relatively higher BWR than the pure working fluids. 0.67R245fa/0.33R123 owns the highest BWR, while R123 yields the lowest BWR. BWR is in range of 11.86–23.22%, indicating that improving the pump operation characteristics is one way to enhance the ORC performance.

Figure 12 shows the expander shaft power with mass flow rate for R245fa, R123 and their mixtures. It can be seen that as the mass flow rate increases, the expander shaft power for R245fa, 0.67R245fa/0.33R123 and 0.33R245fa/0.67R123 present a

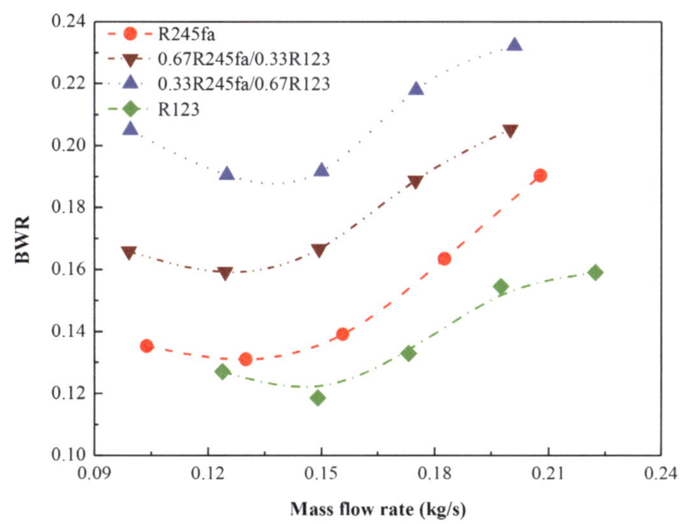

Figure 11.
Variation of BWR with mass flow rate for R245fa, R123 and their mixtures.

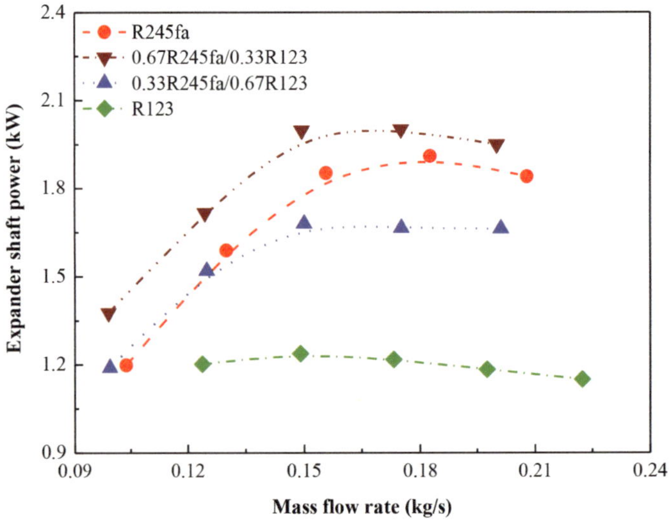

Figure 12.
Variation of expander shaft power with mass flow rate for R245fa, R123 and their mixtures.

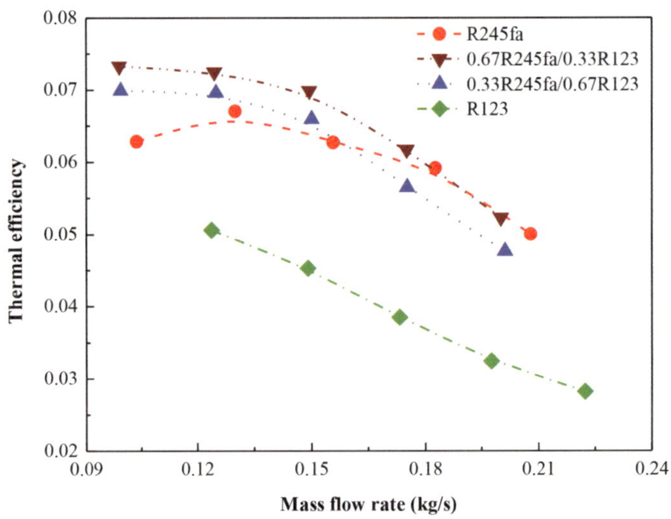

Figure 13.
Variation of thermal efficiency with mass flow rate for R245fa, R123 and their mixtures.

similar trend of increase first and then decrease, while that of R123 almost has no change. The highest expander shaft power is obtained by 0.67R245fa/0.33R123, while the lowest one is got by R123. One optimum mass flow rate is existed to ensure the highest expander shaft power. The maximum expander shaft power for R245fa of 2.76 kW, 0.67R245fa/0.33R123 of 2.85 kW, 0.33R245fa/0.67R12 of 2.46 kW and R123 of 1.82 kW are obtained.

The thermal efficiencies for R245fa, R123 and their mixtures are plotted in **Figure 13**. As observed, as the mass flow rate increases, the thermal efficiencies for 0.33R245fa/0.67R123, 0.67R245fa/0.33R123 and R123 keep decreasing, whereas that of R245 presents a parabolic trend. The thermal efficiency is determined by the net power output and heat input. The main reason is the comprehensive effect of the increasing net power output and heat input. It also can be found that the mixture working fluids have a relatively higher thermal efficiency than the

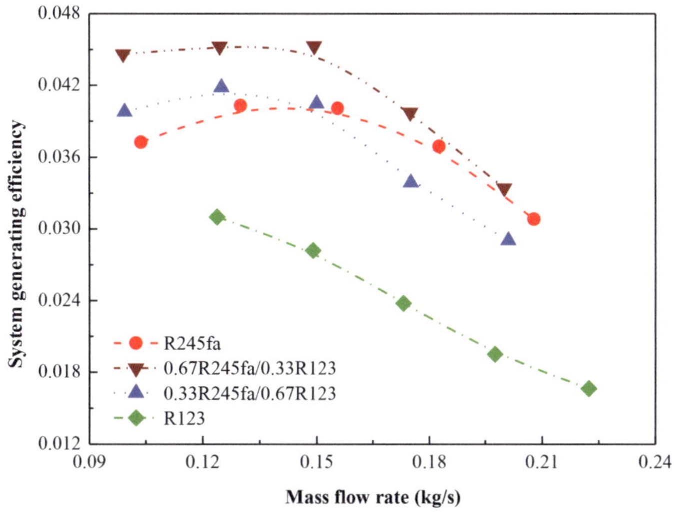

Figure 14.
Variation of system generating efficiency with mass flow rate for R245fa, R123 and their mixtures.

pure working fluids. The maximum thermal efficiencies for R245fa of 6.70%, 0.67R245fa/0.33R123 of 7.33%, 0.33R245fa/0.67R123 of 6.99% and R123 of 5.06% are obtained. 0.67R245fa/0.33R123 owns the highest maximum thermal efficiency of 7.33, 9.4% higher than that of R245fa and 44.86% higher than that of R123.

Figure 14 shows the system generating efficiencies with mass flow rate for R245fa, R123 and their mixtures. The system generating efficiencies for different working fluids have a same trend with the thermal efficiency (as shown in **Figure 13**). The comprehensive influence of heat input and net electricity output enables the parabolic trend in system generating efficiency. The maximum system generating efficiencies for R245fa of 4.03%, 0.67R245fa/0.33R123 of 4.53%, 0.33R245fa/0.67R123 of 4.18% and R123 of 3.10% are yielded. Meanwhile, the corresponding mass flow rate for maximum system generating efficiencies of R245fa, 0.67R245fa/0.33R123, 0.33R245fa/0.67R123 and R123 are 0.130, 0.149, 0.125 and 0.124 kg/s, respectively. 0.67R245fa/0.33R123 owns the highest system generating efficiency of 4.53%, which is 12.41% higher than that of R245fa.

Through the experimental comparison between the pure and mixture working fluids, it indicates that whether the mixtures exhibit better thermodynamic performance than the pure working fluids depend on the operation parameters and mass fraction of mixtures. Meanwhile, the mixture working fluids obtain a higher expander shaft power but a relatively higher BWR, indicating enhancement in pump is necessary for ORC application.

6. System behaviors at different operation strategies

To ensure that ORC can operate at different zone, two operation strategies are proposed: stand alone and gird connect operation strategies. The mass flow rates and expander rotating speed with heat input for standalone and gird connect operation strategies are displayed in **Figure 15**. As the heat input increases, the mass flow rate for standalone operation strategy is in range of 0.153–0.359 kg/s, while that of grid connect operation strategy increases from 0.230 to 0.420 kg/s. Meanwhile, the expander rotating speed for standalone operation strategy keeps rising from 2320 to 2983 rpm, whereas that of grid connect operation strategy keeps constant of 3600 rpm.

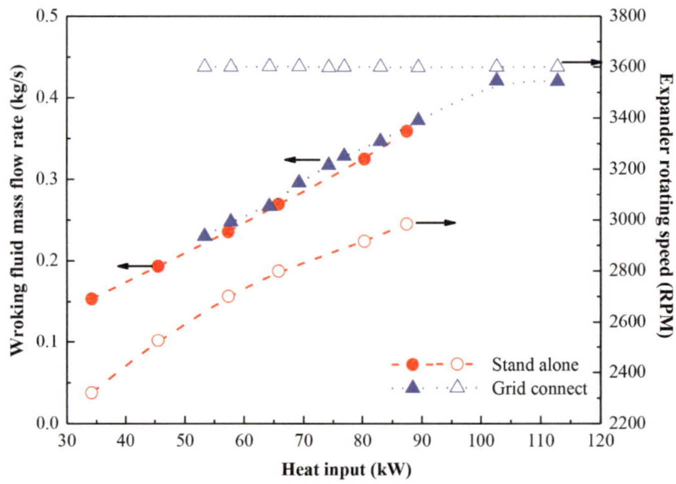

Figure 15.
Variation of working fluid mass flow rates and expander rotating speed with heat input for standalone and gird connect operation strategies.

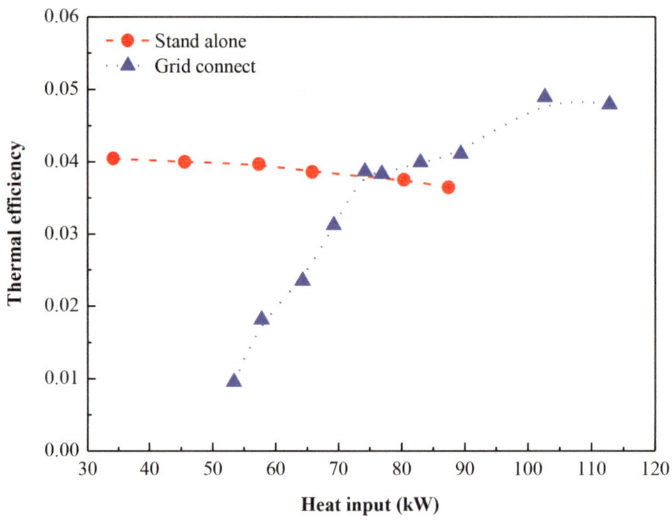

Figure 16.
Variation of thermal efficiencies with heat input for standalone and gird connect operation strategies.

Figure 16 shows the thermal efficiencies with heat input for standalone and gird connect operation strategies. When the heat input increases from 34.18 to 87.40 kW, the thermal efficiency for standalone operation strategy decreases 4.04–3.64%, which may be attributed to the comprehensive effect of increasing heat input. However, for the gird connect operation strategy, the thermal efficiency increases from 0.96–4.89% with heat input. Meanwhile, when the heat input is lower than 74.20 kW, the stand alone operation strategy owns a relatively higher thermal efficiency than the gird connect operation strategy, but an opposite trend for the heat input higher than 74.20 kW. When the heat input is 102.62 kW, the gird connect operation strategy yields the highest thermal efficiency of 4.89%.The system generating efficiencies with heat input for standalone and gird connect operation strategies are demonstrated in **Figure 17**. The system generating efficiencies for standalone and gird connect operation strategies present a similar trend with the thermal efficiencies (as shown in **Figure 16**). The maximum system generating

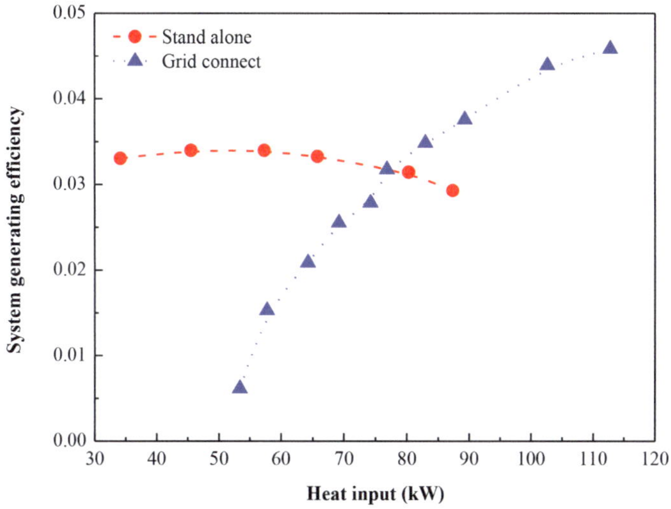

Figure 17.
Variation of system generating efficiencies with heat input for standalone and gird connect operation strategies.

efficiency for gird connect operation strategy is 4.59%, which is 35% higher than that of standalone operation strategy approaching of 3.40%.

7. Conclusion

Power conversion systems based on organic Rankine cycles have been identified as a potential technology especially in converting low-grade waste heat into electricity. This chapter presents the detailed operation characteristic of a small-scale ORC. The effect of mass flow rate, pressure drop, degree of superheating and condenser temperature on thermal efficiency and system generating efficiency are examined. Two pure working fluids (R245fa, R123) and two mixtures working fluids (0.67R245fa/0.33R123 and 0.33R245fa/0.67R123) are tested and compared. The system behaviors at two different operation strategies are addressed. The pressure drop has a relatively effect on the system performance, indicating that improving the system overall performance should give priority to increase the pressure drop. The maximum thermal efficiency does not represent the highest electrical power, which is in line with the theoretical study. The degree of superheating exhibits insensitive on the electrical power. There is an optimum mass flow rate to ensure the minimum BWR. Whether the mixtures exhibit better thermodynamic performance than the pure working fluids depend on the operation parameters and mass fraction of mixtures. Meanwhile, the mixture working fluids obtain a higher expander shaft power but a relatively higher BWR. The expander rotating speed for standalone operation strategy keeps rising from 2320 to 2983 rpm, whereas that of grid connect operation strategy keeps constant of 3600 rpm. When the heat input is lower than 74.20 kW, the stand alone operation strategy owns a relatively higher thermal efficiency than the gird connect operation strategy, but an opposite trend for the heat input higher than 74.20 kW.

Acknowledgements

This research work has been supported by the Ministry of Science and Technology, Taiwan under the grants of Contract No. MOST 107-2221-E-027-091, and by "Research Center of Energy Conservation for New Generation of Residential,

Commercial, and Industrial Sector" from The Featured Areas Research Center Program within the framework of the Higher Education Sprout Project by the Ministry of Education (MOE) in Taiwan. The authors are grateful for National Natural Science Foundation of China (51806081), the Natural Science Foundation of Jiangsu Province (BK20180882), the China Postdoctoral Science Foundation (2018 M632241) and the Open Foundation Program of Key Laboratory of Efficient Utilization of Low and Medium Grade Energy (Tianjin University), the Ministry of Education of China (201806-402).

Conflict of interest

The author declared that there is no conflict of interest.

Author details

Tzu-Chen Hung[1*] and Yong-qiang Feng[2]

1 Department of Mechanical Engineering, National Taipei University of Technology, Taipei, Taiwan

2 School of Energy and Power Engineering, Jiangsu University, Zhenjiang, China

*Address all correspondence to: tchung@ntut.edu.tw

References

[1] Feng YQ, Hung TC, Zhang Y, Li BX, Yang JF, Shi Y. Performance comparison of low-grade organic Rankine cycles (ORCs) using R245fa, pentane and their mixtures based on the thermoeconomic multi-objective optimization and decision makings. Energy. 2015;**99**:2018-2029. DOI: 10.1016/j.energy.2015.10.065

[2] Feng YQ, Zhang YN, Li BX, Yang JF, Shi Y. Comparison between regenerative organic Rankine cycle (RORC) and basic organic Rankine cycle (BORC) based on thermoeconomic multi-objective optimization considering exergy efficiency and levelized energy cost (LEC). Energy Conversion and Management. 2015;**96**:58-71. DOI: 10.1016/j.enconman.2015.02.045

[3] Feng YQ, Hung TC, Greg K, Zhang YN, Li BX, Yang JF. Thermoeconomic comparison between pure and mixture working fluids for low-grade organic Rankine cycles (ORCs). Energy Conversion and Management. 2015;**106**:859-872. DOI: 10.1016/j.enconman.2015.09.042

[4] Tian H, Shu GQ, Wei HQ, Liang XY, Liu LN. Fluids and parameters optimization for the organic Rankine cycles (ORCs) used in exhaust heat recovery of internal combustion engine (ICE). Energy. 2012;**47**:125-136. DOI: 10.1016/j.energy.2012.09.021

[5] Tchanche B, Papadakis G, Lambrinos G, Frangoudakis A. Fluid selection for a low-temperature solar organic Rankine cycle. Applied Thermal Engineering. 2009;**29**:2468-2476. DOI: 10.1016/j.applthermaleng.2008.12.025

[6] Dong BS, Xu GQ, Luo X, Zhuang LH, Quan YK. Potential of low temperature organic Rankine cycle with zeotropic mixtures as working fluid. Energy Procedia. 2017;**105**:1489-1494. DOI: 10.1016/j.egypro.2017.03.444

[7] Chang JC, Hung TC, He YL, Zhang WP. Experimental study on low-temperature organic Rankine cycle utilizing scroll type expander. Applied Energy. 2015;**155**:150-159. DOI: 10.1016/j.apenergy.2015.05.118

[8] Muhammad U, Imran M, Lee DH, Park BS. Design and experimental investigation of a 1 kW organic Rankine cycle system using R245fa as working fluid for low-grade waste heat recovery from steam. Energy Conversion and Management. 2015;**103**:1089-1100. DOI: 10.1016/j.enconman.2015.07.045

[9] Mathias JA, Johnston JR, Cao JM, Priedeman DK, Christensen RN. Experimental testing of gerotor and scroll expanders used in, and energetic and exergetic modeling of, an organic Rankine cycle. Journal of Energy Resources Technology Transactions ASME. 2009;**131**:21-24. DOI: 10.1115/1.3066345

[10] Carraro G, Pallis P, Leontaritis AD, Karellas S, Vourliotis P, Rech S, et al. Experimental performance evaluation of a multi-diaphragm pump of a micro-ORC system. Energy Procedia. 2017;**129**:1018-1025. DOI: 10.1016/j.egypro.2017.09.232

[11] Yang XF, Xu JL, Miao Z, Zou JH, Yu C. Operation of an organic Rankine cycle dependent on pumping flow rates and expander torques. Energy. 2015;**90**:1-15. DOI: 10.1016/j.energy.2015.07.121

[12] Xu W, Zhang JY, Zhao L, Deng S, Zhang Y. Novel experimental research on the compression process in organic Rankine cycle (ORC). Energy Conversion and Management. 2017;**137**:1-11. DOI: 10.1016/j.enconman.2017.01.025

[13] Lei B, Wang JF, Wu YT, Ma CF, Wang W, Zhang L, et al. Experimental

study and theoretical analysis of a Roto-Jet pump in small scale organic Rankine cycles. Energy Conversion and Management. 2016;**111**:198-204. DOI: 10.1016/j.enconman.2015.12.062

[14] Zeleny Z, Vodicka V, Novotny V, Mascuch J. Gear pump for low power output ORC—An efficiency analysis. Energy Procedia. 2017;**129**:1002-1009. DOI: 10.1016/j.egypro.2017.09.227

[15] Bianchi G, Fatigati F, Murgia S, Cipollone R, Contaldi G. Modeling and experimental activities on a small-scale sliding vane pump for ORC-based waste heat recovery applications. Energy Procedia. 2016;**101**:1240-1247. DOI: 10.1016/j.egypro.2016.11.139

[16] Wu TM, Liu JH, Zhang L, Xu XJ. Experimental study on multi-stage gas-liquid booster pump for working fluid pressurization. Applied Thermal Engineering. 2017;**126**:9-16. DOI: 10.1016/j.applthermaleng.2017.07.159

[17] Yang YX, Zhang HG, Xu YH, Yang FB, Wu YT, Lei B. Matching and operating characteristics of working fluid pumps with organic Rankine cycle system. Applied Thermal Engineering. 2018;**142**:622-631. DOI: 10.1016/j.applthermaleng.2018.07.039

[18] Landelle A, Tauveron N, Revellin R, Haberschill P, Colasson S, Roussel V. Performance investigation of reciprocating pump running with organic fluid for organic Rankine cycle. Applied Thermal Engineering. 2017;**113**:962-969. DOI: 10.1016/j.applthermaleng.2016.11.096

[19] Meng FX, Zhang HG, Yang FB, Hou XC, Lei B, Zhang L. Study of efficiency of a multistage centrifugal pump used in engine waste heat recovery application. Applied Thermal Engineering. 2017;**110**:779-786. DOI: 10.1016/j.applthermaleng.2016.08.226

[20] Bianchi M, Branchini L, Casari N, De PA, Melino F, Ottaviano S, et al. Experimental analysis of a micro-ORC driven by piston expander for lowgrade heat recovery. Applied Thermal Engineering. 2019;**148**:1278-1291. DOI: 10.1016/j.applthermaleng.2018.12.019

[21] Yang YX, Zhang HG, Xu YH, Zhao R, Hou XC, Liu Y. Experimental study and performance analysis of a hydraulic diaphragm metering pump used in organic Rankine cycle system. Applied Thermal Engineering. 2018;**132**:605-612. DOI: 10.1016/j.applthermaleng.2018.01.001

[22] Bianchi G, Fatigati F, Murgia S, Cipollone R. Design and analysis of a sliding vane pump for waste heat to power conversion systems using organic fluids. Applied Thermal Engineering. 2017;**124**:1038-1048. DOI: 10.1016/j.applthermaleng.2017.06.083

[23] Sun HC, Qin J, Hung TC, Hua HY, Ya PG, Lin CH. Effect of flow losses in heat exchangers on the performance of organic Rankine cycle. Energy. 2019;**172**:391-400. DOI: 10.1016/j.energy.2019.01.131

[24] Xi H, Zhang HH, He YL, Huang ZH. Sensitivity analysis of operation parameters on the system performance of organic Rankine cycle system using orthogonal experiment. Energy. 2019;**172**:435-442. DOI: 10.1016/j.energy.2019.01.072

[25] Borsukiewicz-Gozdur A. Pumping work in the organic Rankine cycle. Applied Thermal Engineering. 2013;**51**:781-786. DOI: 10.1016/j.applthermaleng.2012.10.033

[26] Dong BS, Xu GQ, Cai Y, Li HW. Analysis of zeotropic mixtures used in high-temperature organic Rankine cycle. Energy Conversion and Management. 2014;**84**:253-260. DOI: 10.1016/j.enconman.2014.04.026

[27] Garg P, Kumar P, Srinivasan K, Dutta P. Evaluation of isopentane, R-245fa and their mixtures as working fluids for organic Rankine cycles. Applied Thermal Engineering. 2013;**51**(1):292-300. DOI: 10.1016/j.applthermaleng.2012.08.056

[28] Lecompte S, Ameel B, Ziviani D, van den Broek M, De Paepe M. Exergy analysis of zeotropic mixtures as working fluids in organic Rankine cycles. Energy Conversion and Management. 2014;**85**:727-739. DOI: 10.1016/j.enconman.2014.02.028

[29] Zhao L, Bao JJ. The influence of composition shift on organic Rankine cycle (ORC) with zeotropic mixtures. Energy Conversion and Management. 2014;**83**:203-211. DOI: 10.1016/j.enconman.2014.03.072

[30] Liu Q, Duan Y, Yang Z. Effect of condensation temperature glide on the performance of organic Rankine cycles with zeotropic mixture working fluids. Applied Energy. 2014;**115**:394-404. DOI: 10.1016/j.apenergy.2013.11.036

[31] Pu W, Yue C, Han D, He W, Liu X, Zhang Q, et al. Experimental study on organic Rankine cycle for low grade thermal energy recovery. Applied Thermal Engineering. 2016;**94**:221-227. DOI: 10.1016/j.applthermaleng.2015.09.120

[32] Molés F, Navarro-Esbrí J, Peris B, Mota-Babiloni A. Experimental evaluation of HCFO-1233zd-E as HFC-245fa replacement in an organic Rankine cycle system for low temperature heat sources. Applied Thermal Engineering. 2016;**98**:954-961. DOI: 10.1016/j.applthermaleng.2016.01.011

[33] Jung HC, Taylor L, Krumdieck S. An experimental and modelling study of a 1 kW organic Rankine cycle unit with mixture working fluid. Energy. 2015;**81**:601-614. DOI: 10.1016/j.energy.2015.01.003

[34] Li T, Zhu J, Fu W, Hu K. Experimental comparison of R245fa and R245fa/R601a for organic Rankine cycle using scroll expander. International Journal of Energy Research. 2015;**39**(2):202-214. DOI: 10.1002/er.3228

[35] Feng YQ, Hung TC, Wu SL, Lin CH, Li BX, Huang KC, et al. Operation characteristic of a R123-based organic Rankine cycle depending on working fluid mass flow rates and heat source temperature. Energy Conversion and Management. 2017;**131**:55-68. DOI: 10.1016/j.enconman.2016.11.004

[36] Yang SC, Hung TC, Feng YQ, Wu CJ, Wong KW, Huang KC. Experimental investigation on a 3 kW organic Rankine cycle for low grade waste heat under different operation parameters. Applied Thermal Engineering. 2017;**113**:756-764. DOI: 10.1016/j.applthermaleng.2016.11.032

[37] Feng YQ, Hung TC, He YL, Wang Q, Wang S, Li BX, et al. Operation characteristic and performance comparison of organic Rankine cycle (ORC) for low-grade waste heat using R245fa, R123 and their mixtures. Energy Conversion and Management. 2017;**144**:153-163. DOI: 10.1016/j.enconman.2017.04.048

[38] Feng YQ, Hung TC, Su TY, Wang S, Wang Q, Yang SC, et al. Experimental investigation of a R245fa-based organic Rankine cycle adapting two operation strategies: Stand alone and grid connect. Energy. 2017;**141**:1239-1253. DOI: 10.1016/j.energy.2017.09.119

[39] NIST Standard Reference Database 23. NIST thermodynamic and transport properties of refrigerants and refrigerant mixtures REFROP, Version 9.0. 2010

Chapter 5

A Recent Review in Performance of Organic Rankine Cycle (ORC)

Syamimi Saadon and Salmah Md Saiful Islam

Abstract

Increasing emissions of carbon dioxide and fuel prices lead to extra efforts in finding solution to reduce the environment waste heat. One of the solutions emerging is the organic Rankine cycle (ORC) system. It is one of the promising exhaust heat recovery technologies which is widely been used to recover low to medium-grade heat rather than conventional steam Rankine cycle system. This chapter highlights on the different conditions and configurations of ORCs that are usually been applied. These different configurations have different constraints and usually will be considered based on the applications.

Keywords: organic Rankine cycle, preheater, supercritical, superheating, waste heat recovery

1. Introduction

As higher efficiency of industrial technology is in demand, more and latest technologies are needed to produce energy. Increasing of population growth [1] and escalating process of electricity are mostly due to emission gases from the industry, vehicles, deforestation and others. In the aerospace industry, engineers continuously search for new methods to upgrade the efficiency of the engines. Recovery waste heat could increase the engine efficiency [2]. Although society undergoes global issues, social problems or economy crisis, this does not stop aerospace industry from expanding which leads to increase in demand on aircraft. This results in increasing of fuel price since more conventional fuel is needed and causes pollution into the environment [3]. Still, the price in this development is used to optimize the engine. In each flight, greater revenue could be achieved when the number of passengers is greater. And to have more passengers, lighter aircraft is needed. It is important to note that weight is very crucial in changing aircraft engines as it connects linearly with the amount of fuel used in every unit of force powered by engine (specific fuel consumption). Waste heat recovery (WHR) is one of the most important solutions found to lower the emission and fuel consumption [4]. Waste heat or low- grade waste heat is heat energy produced in the atmosphere through internal combustion. However, these low-grade waste heat are not sufficient enough to generate power due to insufficient low temperature. Thus, to recover these waste heat, organic Rankine cycle (ORC) system is one of the beneficial exhaust heat recovery technologies which is widely utilized in the applications of low grade heat recovery rather than conventional Rankine cycle [5]. By combining an ORC with energy system, for instance in power plants, organic fluid of low boiling point is utilized to change heat into electricity. The organic fluids or refrigerants used in air conditioning systems accumulates (collect) heat from a volume of air and release it to different type of heat exchanger which increases the expansion of high vapor pressure in expander. The heat accumulated is transformed into mechanical power or electricity and therefore will help to increase the thermal efficiency and the overall performance of the engine. Thus, higher thrust could be obtained as less electrical power is needed from aircraft engine resulting lower engine bleed air [2, 5, 6]. Because of its thermodynamic properties, organic fluid is the best selection for low quality heat sources with temperatures below than 100°C [2]. By selecting proper working fluid for low waste heat recovery system and modeling an optimum design of heat exchanger configurations, the waste energy recovered through this ORC system could be maximized. Thus, designing a fuel-efficient and cheaper heat exchanger, ORC power plant can effectually utilize the economic and environmental issues especially in aerospace industry.

2. Aspects of recovery waste heat systems

Based on second law of thermodynamics, the efficiency of a process would not be 100% as there is no process that can entirely transform all amount of heat into work. The energy that is not used to produce work is being dissipated as heat at different temperatures, levels streams. On aircraft, half of the fuel energy lost through this way. Nevertheless, these sources of waste heat are everywhere; from this lost energy, only a part of it can be used to produce mechanical work or other purposes, where around 30% of the total waste heat could be changed to useful work. As the demand of aircraft is increasing vastly, the aviation industry has been the world center of attraction as new technologies are needed and maximum exploitation of fuel is a must. The conversion of heat energy to mechanical or electrical power depends on the characteristics of the source. Let say in an air conditioning system, an external hose is two or three degrees above the ambient temperature, it is a waste to recover that little amount of energy, however, this power leak will be an irreversibility process together with other similar leaks will decrease the thermal efficiency. This is called as waste heat and is an unused heat energy produced as a by-product of process of energy transformation, as a natural consequence on any non-adiabatic process from the thermodynamics law. Most of the available waste heat is low waste heat that can be used by an ORC which utilizes low boiling point organic fluid as working fluid, for example, toluene, hexane or pentane. Presently, there has not been any waste heat recovery (WHR) system added to an aircraft. Nevertheless, researchers suggest on adding WHR system to future engines and propose to make changes in current engines. However, it is a hassle to change the actual design of the engine as more expenses will be used in research, tests and certifications and a lot of heat source needs to be taken into account. Pasini et al. [7] analyzed the possibilities of heat recovery results in overall efficiency of an aircraft engine. A waste heat recovery system is modeled in a jet engine and a turbo propeller engine. Their project takes into account the nozzle works in off design state. The heat emitted influences greatly in the system performance. They also developed a numerical thermodynamic code to evaluate the positive impacts of waste heat recovery in a turboprop, a turbofan and a turbojet. The turbofan engine is of great interest due to large fraction of thrust is provided by cold flow, whilst gas generator supplies needed power. The authors then concluded that the enthalpy level ahead of exhaust nozzle of gas generator could be decreased without losing a lot of thrust. From the results of the calculations, it was found an increase of thermal efficiency about

4% when heat recovery was done (efficiency of regeneration was 0.5). At the same time, if the efficiency was 0.7, an increase of 10% was achieved. From the numerical simulations, the best place for heat recovery is from hot gas before entering nozzle. Another research done by Li et al. [8] was to study the small-scale ORC system performance with low grade heat sources to provide electricity in various working state. The experiment setup includes normal ORC system components, for instance, turboexpander with high speed generator, finned tube condenser, ORC pump and plate evaporator. Results found that the turbine power and condenser heat output, ORC pump power and evaporator heat input, turbine isentropic, overall efficiencies and system thermal efficiencies rises when heat source temperature rises too. The fluid of ORC during superheat and pressure at turbine inlet were two crucial variables that kept constant with temperature heat source and pump speed of ORC.

2.1 Organic Rankine cycle

ORC utilizes organic compound instead of water as a working fluid, generally, a refrigerant, a hydrocarbon such as pentane, butane, perfluorocarbon or silicon oil. The organic fluid's boiling point is much lesser compared to water and enable heat recovering at lesser temperatures instead of the steam Rankine cycle [9]. ORC's first commercial applications with medium-scale power plants for geothermal and solar applications were developed in the 9 late 1970s and 1980s. These days, over 200 ORC power plants are recognized with more than 1800 MWe installed and the technologies keep on increasing day by day [10]. Mostly, the plants were installed in biomass CHP application, geothermal plants and plants of WHR followed. The layout of ORC is much simpler compared to the steam cycle as there is no water vapor attached to the boiler, and a single heat exchanger could be utilized for the three processes of evaporation including preheating, vaporizing and superheating. ORC is able to use low grade heat sources than steam Rankine cycle. Since it could be utilized in lower temperature at the turbine inlet and reduce thermal stresses in the boiler. In regular steam plant systems, the performance cycle is at risk damage due to gaseous infiltrations that occur in sub atmospheric condensing pressure. In steam-based cycle, the usage of a single tube for evaporation is abstained due to large density difference that exists in between liquid and vapor phases. However, some organic fluids have condensation pressure higher than the atmospheric pressures and this avoids the infiltration of non-condensable gases in the condenser. The small differences in density organic fluid phase of liquid and vapor also enables the use of once-through boilers. This led to avoidance of using stream drums and simplified the operation of the whole plant. A simple plant system can be developed and less cost is needed when uses organic fluid compared to steam based cycle. In ORC, usage of deaerator is unnecessary but that is not the case for steam base cycle. Due to presence of oxygen, water deaerator or water treatment must be added to avert erosion. Because of low fluid density in the cycle low-pressure part, steam cycle also needs large turbines, heat exchangers and hydraulic diameter for pipes. Meanwhile, since organic fluid has higher density fluid, usage of compact appliances is allowed, especially in marine application, the available space for recovery plant of waste heat is restricted. Other than that, the enthalpy drop in ORC is much lower compared to steam cycle. The process in ORC can be done in a single stage with much simpler turbine compared to steam cycle which requires turbine with some expansion stages. ORC normally operated at much lower pressure levels and rarely exceeding 30 bars. Thus, ORC is beneficial in low to medium power range due to its cycle simplicity, less cost and stress level needed at boiler, easier to control and simpler usage of components [11].

2.1.1 Difference between steam Rankine cycle and organic Rankine cycle

Traditional steam Rankine Cycle utilizes water and higher pressure vapor as the main flow fluid in the cycle and mainly used in high temperature more than 500°C. However, ORC uses organic pure less boiling point or working fluid mixture mostly used in lower temperature process less than 500°C. Thus, ORC is more advantageous in recovering less temperature heat energy for various temperature ranges. Although steam Rankine cycle is the most common technology used in recovering heat process that converts heat into electricity, it is unfit for conditions of low temperature and pressure. This results from the need for high temperature and pressure in operation. If exhaust temperature and pressure are similar, the SRC exhaust steam enthalpy is greater and the heat from the cold sources increases [12].

Based on the **Figure 1** which is the T-s diagram above, two important differences could be found. Firstly, the curve of the organic fluid is abundantly vertical, meanwhile, water has negative curve of saturated vapor slope. Hence, when process of expansion finish, the limitation of vapor quality invisibles in ORC cycle and superheating vapor is unnecessary before the turbine inlet. Second, the gap in entropy between saturated liquid and saturated vapor for organic fluids is lesser. Thus, the enthalpy vaporization is lesser. The organic working fluid mass flow rate should therefore be greater than water to absorb equal thermal power in the evaporator, resulting in more pump consumptions [13].

2.1.2 Supercritical organic Rankine cycle

ORC can be operated in a subcritical or supercritical cycle. In supercritical cycle, the working fluid evaporation ends in supercritical area and the heat rejection in condenser occurs in the subcritical area. Many studies have been performed on the supercritical ORC. **Figure 2** provides the temperature and entropy changes of supercritical ORC.

Yagli et al. [14] modeled subcritical and supercritical ORC to recuperate exhaust gas waste heat of biogas fuelled CHP engine. Comparing with subcritical condition,

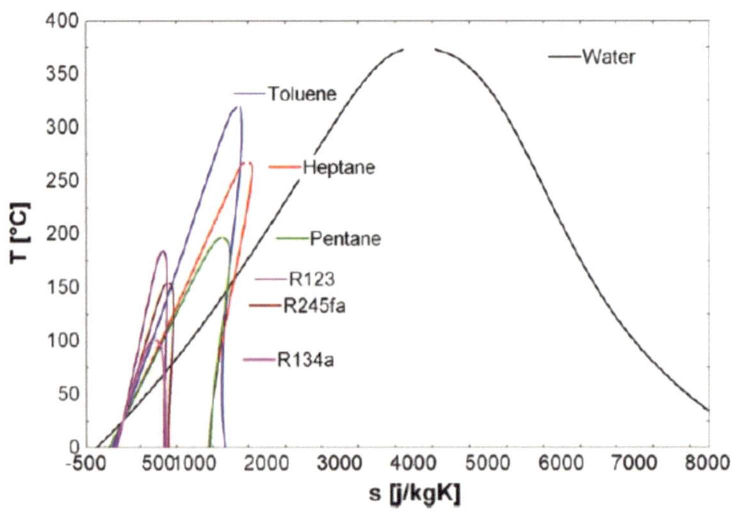

Figure 1.
T-s diagram of the water saturation curves of water and a few typical ORC organic fluids applications [13].

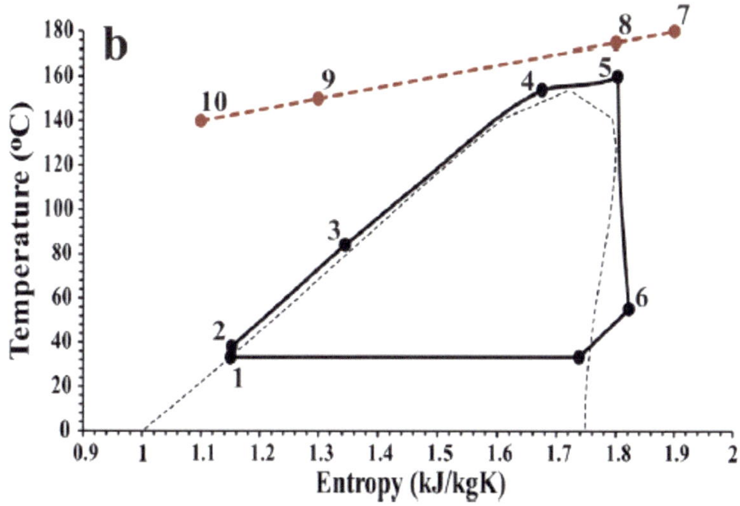

Figure 2.
Supercritical ORC temperature and entropy change [14].

supercritical ORC shows greater performance. At constant pressure, supercritical ORC performance rise as turbine inlet temperature rises. The most excellent performed cycle net power, thermal efficiency and exergy efficiency are evaluated as 79.23 kW, 15.51 and 27.20% for subcritical and 81.52 kW, 15.93 and 27.76% for supercritical ORC, respectively. Guo et al. [15] studied the subcritical and transcritical ORC performance in regards to the evaporator pinch point locations. Found that transcritical ORCs gives higher performance as the heat source outlet temperatures lessen. Ran et al. [16] studied the impact of big transformation in the thermophysical properties of pseudocritical region. Utilizing network output, thermal efficiency and total vapor area an optimization method was found. The results showed that in transcritical ORC's, the thermophysical properties of the working fluid work at supercritical coefficient and logarithmic mean temperature difference (LMTD). Moloney et al. [17] analyzed the pressure effect to optimize the first law efficiency, second law's efficiency and net power of a supercritical ORC with 170–240°C turbine inlet temperature suitable for geothermal reservoirs of medium temperature. Found that supercritical cycle is much more efficient than subcritical cycle to optimize the plant efficiency. Chowdury et al. [18] presented an ORC simulation with different source of heat from the actual vehicle exhaust in supercritical state. The simulation shows that the key in transforming the operating temperature at the evaporator outlet is to modify the mass flow rate.

2.1.3 Application of organic Rankine cycle

The ORC technology has been utilized broadly and applied in various industrial activities especially in biomass and geothermal application. Nevertheless, ORC technology has been increasing in solar thermal system and heat recovery applications from industrial waste heat.

2.1.3.1 The combined heat and power (CHP) of biomass

There is widespread use of agricultural or industrial processes such as lumber or agricultural waste in biomass due to low energy density than the fossil fuels and availability of heat and electricity, where biomass is suitable on off-grid case or unreliable grid connection. Local generation results in smaller power plants that exclude traditional steam cycles that in this power range are not profit-making.

Figures 3 and **4** define the working principle of such cogeneration system, at a temperature from 150 to 320°C, heats from combustion is transmitted from the flue gases to the heat transfer fluid in two heat exchangers. When temperature lowers a little bit below 300°C, heat transfer fluid (thermal oil) is sent to the ORC loop to evaporate the working fluid. Then, the evaporated fluid expands, to preheat the liquid using recuperator and when temperature reached 90°C, the fluid condensed to produce hot water.

ORC efficiency is lesser compared to traditional steam cycles and gradually reduces for small scale units. To raise the overall energy conversion efficiency of plant, heat demand is needed and could be met through space heating or industrial processes (wood drying). Load of plant could be managed through on- site heat request or maximize power generation which includes additional wasting heats but increases, the full load operating hours per year.

From **Figure 3**, even though the CHP system's electrical efficiency is somewhat less (18%), the overall system efficiency is 88% greater than centralized power plants where most residual heat is lost. These gases need to be cooled to the least possible value, so that acid dew point could not be achieved and to lower the losses in flue gases. Two heat transfer loops are utilized to achieve this point (high and low temperature). The lower temperature loops are installed after the high flue temperature to lower the outlet temperature. Competitive technology in generating electric out of solid biofuels is biomass gasification where biomass changes into an organic gas mainly consisting H_2, CO, CO_2 and CH_4. In order to remove solid particles, this synthetic gas is treated and filtered and finally burned in an ICE or in a gas turbine. Contrasting Biomass CHP's technology and costs with an ORC with gasification, gasification yields higher investment costs (75%), higher maintenance costs (200%) and more power-to-thermal ratio, where utilization is increase profit-making. ORC is an established technology meanwhile gasification plants are normally used as prototype in operation.

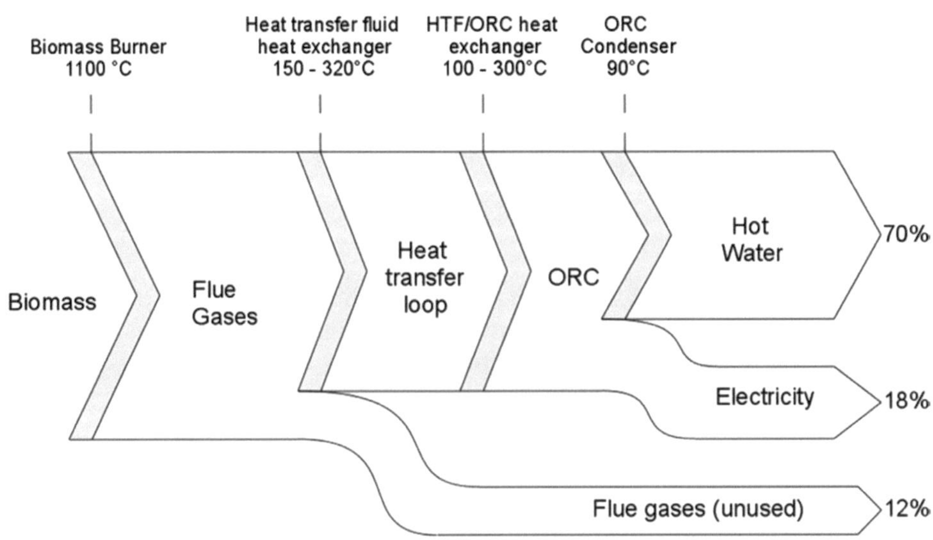

Figure 3.
Energy flows in a CHP system of biomass [13].

2.1.3.2 Geothermal energy

Geothermal heat sources ranges from 10 to 300°C. The actual lower technological limit to generate electricity is about 80°C, and became less efficient with temperature less than 80°C and causes uneconomical geothermal plants. The potential of geothermal energy in Europe is shown in **Table 1** and indicates that low temperature sources have higher potential.

For better production and injection, boreholes need to be drilled in the ground (**Figure 5**) to recover heat at an acceptable temperature. Then, the hot brine is pumped out of the first one and injected at a lower temperature in the second. Boreholes might be few thousand meters deep which results in working continuously for few months depend on the configuration of the geology and causes increasing share of drilling for geothermal plant cost investment (up to 70%) [19]. High auxiliary consumption is also characterized by low geothermal ORC: the pumps ingest 30–50% of the gross power output [20]. The brine pump together with a significant flow rate has to circulate the brine over large stances is the primary consumer. Working fluid of pump consumption is greater than higher temperature cycles, as the ratio of pump consumption to turbine output power

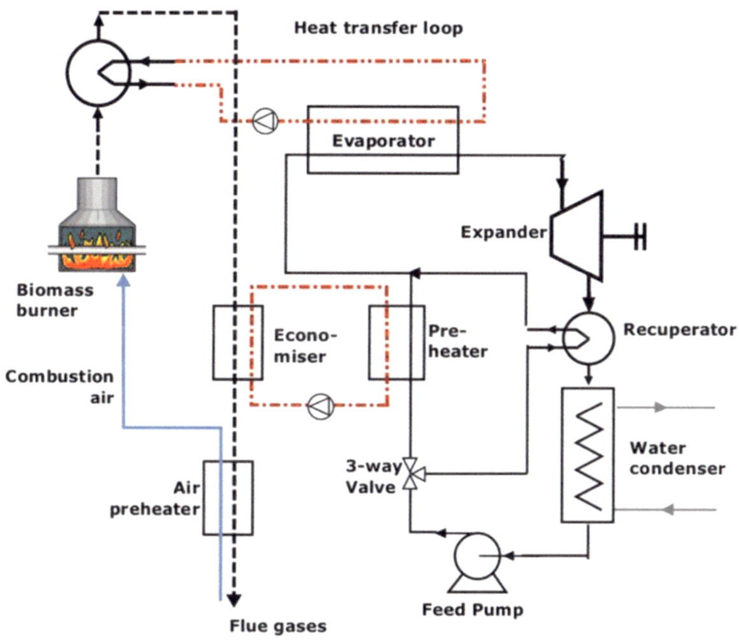

Figure 4.
Biomass CHP ORC system working principle [13].

Temperature	MWth	MWe
65–90°C	147,736	10,462
90–120°C	75,421	7503
120–150°C	22,819	1268
150–225°C	42,703	4745
225–350°C	66,897	11,150

Table 1.
Geothermal energy potential in Europe for different temperature ranges of heat sources [19].

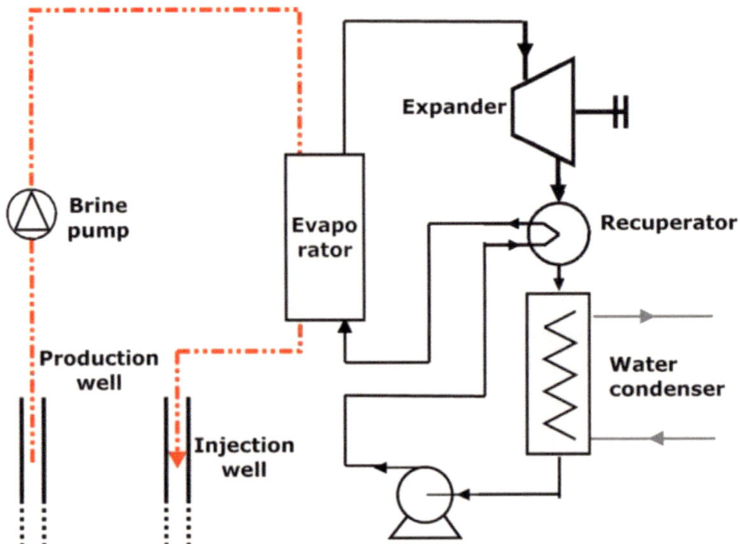

Figure 5.
Geothermal ORC system working principle [13].

('back work ratio') rose as evaporation temperature lowered. Geothermal heat sources temperature (>150°C) allow for CHP, where the condensing temperature is restricted to a higher temperature such as 60°C, enabling district heating uses cooled water. Thus, the overall efficiency of energy recovery rises with lower electrical efficiency expenses.

2.1.3.3 Solar power plants

Solar power concentration is the best technology on a linear or punctual collector that tracks and reflects the sun, transferring heat to high temperature fluid. Electricity is generated as heat is transmitted to a power cycle; electricity is generated. The three primary technologies of concentration are the parabolic platform, solar tower and the parabolic trough. Punctual concentrating technologies consist of parabolic dishes and solar towers, results in more concentration factor and greater temperatures. For solar towers, the Stirling engine (small-scale plants), the steam cycle or even the combined cycle is the best suited power cycles. Parabolic troughs operate at lesser temperature (300–400°C). Till today, they were combined to traditional steam Rankine cycles to generate electricity [21]. Geothermal or biomass power plants for example, steam cycles need higher temperatures, pressure and installed power to be more cost- effective. Organic Rankine cycle is a favorable technology that could lower the small scale of investment costs by working at lesser temperatures and reduce total installed power to kW scale. The working principle of the system is shown in **Figure 6**. As Fresnel linear technology need lower investment costs [22], they are suitable for ORCs but operate at lesser temperature.

Till recently, only a few of CSP plants with ORC are accessible on the market:

- In 2006, at Arizona, a 1MWe solar concentration of ORC power plant was accomplished. The ORC module utilizes n-pentane as the working fluid with 20% efficiency. On design point, the overall solar energy efficiency is 12.1% [23].

- Few small-scale for the applications of remote-off grid were studied. The only proof of concept obtained is that 1KWe system installed for rural

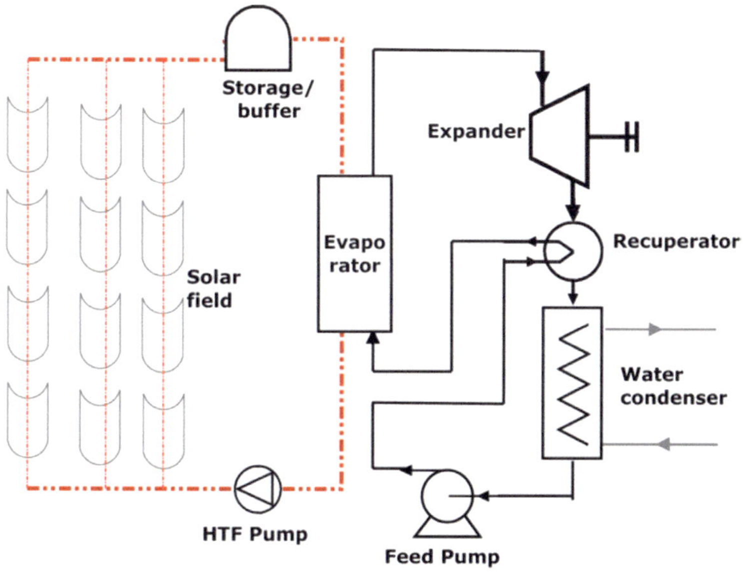

Figure 6.
Solar ORC system working principle [13].

electrification in Lesotho by "STG International". To produce and integrate small size solar thermal technology with medium temperature collectors and an ORC to acquire economics equivalent to big installation of solar thermal is the objective of this project. This design intended to change or adding diesel generators in developing countries at off-grid areas through generating clean power at lesser costs.

2.1.3.4 Mechanical and industrial heat recovery

At low temperature, most of the application in manufacturing industry reject load. Normally, the heat is enormous in large-scale plants, and could not be used again for on-site district heating. The heat then discharged into the atmosphere and results in two types of pollution [24]:

- Health/Environmental issues results from pollutants (CO_2, NO_X, SO_X, HC) of flue gases.

- Unbalance of aquatic equilibrium and negative effector biodiversity due to rejection of heat.

These two types of pollution could be diminished by waste heat recovering. Moreover, it could provide on-site electricity to be consumed or sent it back to the grid. Normally, waste heat is recuperated through an intermediate heat transfer loop in such a system and used to evaporate the cycle's working fluid. In USA, power generation from industrial waste heat sources is approximately about 750 MWe [25]. Some industries have greater potential in recovery of waste heat. One of it, the cement industrial loses 40% of flue gas heat. These flue gases are placed at a temperature of 215–315°C after the preheater of limestone or in the clinker cooler [26]. CO_2 released from the cement industry is 5% of the world's total CO_2 emissions, half of the results from fossil fuels combustion in kilns [24]. Further possible industries include iron and steel industries (for example, 10% of CO_2 emissions in China), refineries or chemical industries. Although their potential is higher and cost-effective (1000–2000 €/kWe), ORC recovery waste heat cycles have only 9–10% of the world's installed ORC plants compared to biomass CHP and geothermal units [10].

2.1.3.5 Aircraft engine

Perullo et al. [27] integrated an ORC to an engine for power generation. They mentioned the problem, as bypass ratio keep on growing and the engine cores becomes effective, the diameter of engine fan increases and the core size decreases which causes pneumatic offset needing greater percentage of the core flow and results in higher performance penalties. They tried to solve the problem by changing the pneumatic off-take to an electrical and used power generated to drive external air to the environmental control system (ECS). With the idea of no-bleed aircraft, performance penalties for shrinking cores and increased fan diameter are supposed to be eliminated and they had demonstrated that a rise in efficiency from 0.9 to 2.5% is possible. Boeing has also applied the no-bleed system; but using generator as the source of energy, not ORC where the generator works with energy taken from the APU and engines. As this application save fuel by 3%, this explain why they put this idea together in ORC rather than extracting energy from fuel, the waste heat could supply the energy needed. ORC is used due to the low quality of the range temperature. The WHR system is placed in the core jet exhaust of a turbofan engine. Conversely to land ORC systems, used in steam power plants for instance, an on-board ORC would suppose operating conditions that may vary continuously in the course of every few hours in external pressure and temperatures. The amount of heat extracted from the engine should be considered to avoid reduction of thrust. The system is distributed in the nozzle, the nose cowl and the Pylon. It uses R245fa as the working fluid having demonstrating highest thermal efficiency in a wide range of operating pressure. The MathCAD 2001 software was used to model the design to govern whether energy is enough to extract our f the exhaust gases to a power of 270 hp. motor. **Figure 7** below describes the ORC schematics.

The model was integrated on a CFM56-7B configuration and cruise conditions were used in it. Some assumptions were done; not analyzing the system with take-off conditions as working fluid dissociates at high temperatures, heat is taken out of core exhaust flow before expanding in the nozzle, assuming a weight of 430 kg and was used to calculate the fuel burn reduction (0.9%) a TSFC reduced its value in 2% compared to the engine alone. It was also assumed that the ORC could produce

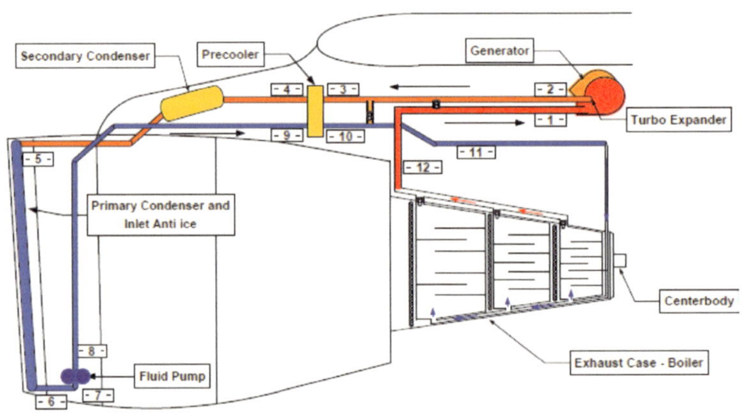

Figure 7.
Aircraft engine ORC system working principle [27].

greater power which is needed to drive the ECS air compressor and resulted in reduction of TSFC for 22%. Perullo et al. [27] concluded that an ORC WHR system could produce more power on the existing engine and can be utilized to supply sufficient power to a compressor driving air to the ECS. They suggested that the design system should be reconfigured to obtain the best results of fuel burn and take into account the need of an electric starting mechanism if the bleed system was removed in future research. The option of using the engine cowl or the anti-icing system in the wing s as the condenser of the ORC system was suggested as well.

2.1.4 Organic Rankine cycle (ORC) with regenerator (RORC)

Regenerative ORCs are designed where ORCs and turbine bleeding are integrated to a heat exchanger. The cycle heats up the working fluid upon infiltrating the evaporator which is almost similar to the ORC with recuperator. **Figure 8(a)** and **(b)** provides the schematic cycle and T-s diagram of regenerative cycle.

At 139°C of turbine inlet temperature, Le et al. [28] utilized a genetic method to optimize the first law and effectiveness of the system for diverse fluids. When examining, CO_2 results the worst while recuperative cycle was discovered for greater efficiencies compared to simple cycle. Moloney et al. [17] studied the environmental fluids with critical temperature below 200°C in regenerative supercritical ORCs to upgrade the geothermal energy efficiency and noted that CO_2 operates the lowest. The same purpose was enforced by Muhammad et al. [29] to the basic ORC; single and double stage regenerative ORC for applications of recovering waste heat. Studies showed that the single and double stage regenerative ORC has greater thermal efficiency with lower economic performance rather than the basic ORC.

2.1.5 Organic Rankine cycle with superheating

Found that superheating of dry fluid negatively affects the ORC's efficiency while wet fluid positively affects the ORC's efficiency and isentropic fluid did not really affect ORC. Nevertheless, an experimental observation by [15] indicated that ORC with wet fluid superheat utilizing R245fa at 1.8°C and if the superheat rises to 8.7°C, the system is stable. Thus, even for dry working fluid, superheating is essential.

Li et al. [30] conducted an experimental study to inquire the performance of a small-scale ORC system with low grade heat source to produce electricity at various

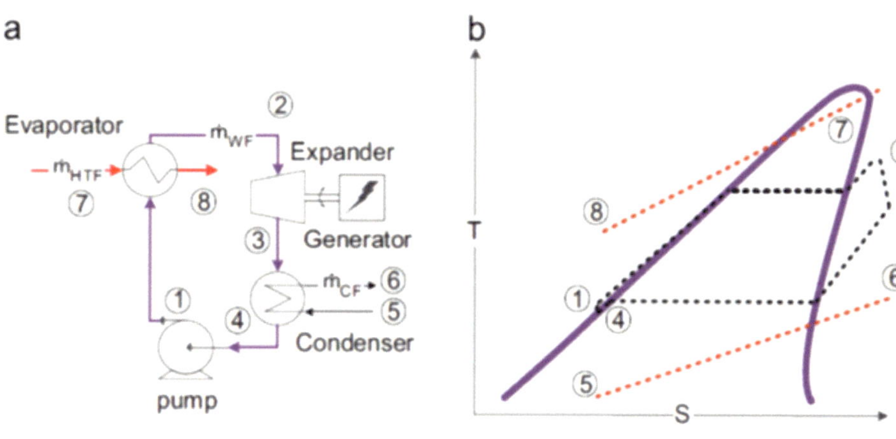

Figure 8.
(a) ORC with regenerator and its (b) T-s diagram [28].

working state. It was found that the fluid of ORC during superheat and pressure at the turbine inlet were two main variables able to be managed with temperature of heat source and speed of the ORC pump. It was also found that superheat and internal heat exchanger are crucial for ORC from both perspectives of thermodynamic and techno-economic. Roy et al. studied the consequences of superheat and recovering on ORC system at certain degree of superheat [31]. Note that Guo et al. [15] argued if the superheat coupled with an internal heat exchanger, greater development could be done. Zhang et al. inquires the consequences of superheat and internal heat exchanger on three ORC designs' thermos-economic performance from fluid properties and heat sources. It has been discovered that the thermo-economic performance of internal heat exchanger ORC with dry surpasses the wet fluid as temperature of heat source load increases [32]. Brizard et al. [33] suggested preventing condensation drops during operation of superheating; the inlet of expander must exceed 20°C. Radulovic et al. [34] mentioned that superheat is important in cycle especially in wet fluids. As the temperature of superheater rises, the cycle efficiency also rises and the chance of the working fluid condenses during pressure drop inside turbine, resulting in corrosion and efficiency drop is lesser. To get a higher efficiencies and net power output, superheating is important to prevent wet expansion. Feng et al. [35] found that rises the superheat degree assure the decrease in mean heat transfer temperature difference in superheating area of evaporator causes decreasing of overall heat transfer area, and decrease in the investment cost of the system. It was also found that outlet temperature of evaporator and superheat degree gives good feedback on the efficiency of exergy. Li et al. [30] construct an investigation on the experimental of a small-scale ORC system under designated working state for the recovery application of low-grade thermal energy. The reaction between condenser cooling water temperatures and superheat.

R245fa at turbine inlet were measured and analyzed on the performance of the system. The outcomes show that when evaporating pressure is constant, superheat at the inlet of expander gives negative feedback on the turboexpander and performance of the at some temperatures of cold water. In conclusion, superheat is crucial in assuring an efficient and safe system operation. Bianchi et al. [36] presented an experimental micro-ORC setup for low-temperature application by implementing a test bench to acquire data for the energy system characterization. From the results, it was found that for the tested working points, efficiency is from 2.9 to 4.4% and increases as degree of superheating decreases. Ismail et al. [37] concluded that utilizing superheated vapor in the system with internal heat exchanger results in increasing of thermal efficiency ORC. The mass flow rate required for the system together with superheated vapor is lower than the saturated vapor system. Thus, superheated is essential to lower the mass flow rate, and enhanced the performance of the system with presence of internal heat exchanger.

3. Conclusion

This chapter presents a comprehensive review on the developments of organic Rankine cycle (ORC) systems that have been used for power generation by using a waste heat source. This review also highlights more on the different configurations of ORCs used, depending on their applications. From here, we could conclude that superheating and the condition of the organic working fluid are crucial in ORC system from the thermodynamic point of views. Therefore, this study plans to investigate the design of an ORC model with better output power by modifying the configurations and adding a superheating device and also to study the effect of using organic dry working fluid (R245fa) at supercritical condition.

Author details

Syamimi Saadon[1,2]* and Salmah Md Saiful Islam[1]

1 Department of Aerospace Engineering, Faculty of Engineering, University Putra Malaysia, Serdang, Malaysia

2 Faculty of Engineering, Aerospace Manufacturing Research Centre (AMRC), University Putra Malaysia, Serdang, Selangor

*Address all correspondence to: mimisaadon@upm.edu.my

References

[1] Rusev TM. Comparative Study of Different Organic Rankine Cycle Models: Simulations and Thermo-Economic Analysis for a Gas Engine Waste Heat Recovery Application. 2014:1-77

[2] Rangel-Martnez D. Assessment of the Integration of an Organic Rankine Cycle for Waste Heat Recovery from Bleed Air for Implementation on Board an Aircraft. Master Thesis. Durham University; 2018

[3] Mazlan NM, Savill M, Kiporous T. Effects of biofuels properties on aircraft. Emerald Insight. 2015;**87**(5):393-401

[4] Baccioli A, Antonelli M. Control variables and strategies for the optimization of a WHR ORC system. Energy Procedia. 2017;**129**:583-590

[5] Patel P, Modh J, Patel V. Parametric Analysis of Organic Rankine Cycle (ORC) for Low Grade Waste Heat Recovery. Proceedings of the National Conference on Thermal Fluid Science and Tribo Application TFSTA2016-50. 2016

[6] Saadon S. Computational modelling of an Organic Rankine Cycle (ORC) waste heat recovery system for an aircraft engine. In: MATEC Web of Conferences, vol. 151, p. 02001. EDP Sciences. 2018:02001

[7] Pasini S, Ghezzi I, Andriani R, Ferri LDA. Heat recovery from aircraft engine. Energy Conversion Engineering Conference and Exhibit, (IECEC), 35th Intersociety. 2000;**6**:8

[8] Li L, Ge YT, Luo X, Tassou SA. Experimental investigations into power generation with low grade waste heat and R245fa organic Rankine cycles (ORCs). Applied Thermal Engineering. 2017;**115**:815-824

[9] Shi L, Shu G, Tian H, Deng S. A review of modified organic Rankine cycles (ORCs) for internal combustion engine waste heat recovery (ICE-WHR). Renewable and Sustainable Energy Reviews. 2018;**92**:95-110

[10] Enertime, "The organic Rankine cycle and its applications. 2011. Available from: http://www.cycle-organique-rankine.com

[11] Sung T, Kim KC. An organic Rankine cycle for two different heat sources: Steam and hot water. Energy Procedia. 2017;**129**:883-890

[12] Zhang X, Wu L, Wang X, Ju G. Comparative study of waste heat steam SRC, ORC and S-ORC power generation systems in medium-low temperature. Applied Thermal Engineering. 2016;**106**:1427-1439

[13] Quoilin S, Van Den Broek M, Declaye S, Dewallef P, Lemort V. Techno-economic survey of organic Rankine cycle system. Renewable and Sustainable Energy Reviews. 2013;**22**:168-186

[14] Yagli H, Koc Y, Koç A, Adnan G, Tandiroglu A. Parametric optimization and exergetic analysis comparison of subcritical and supercritical organic Rankine cycle (ORC) for biogas fuelled combined heat and power (CHP) engine exhaust gas waste heat. Energy. 2016;**111**:923-932

[15] Guo C, Du X, Yang L, Yang Y. Performance analysis of organic Rankine cycle based on location of heat transfer pinch point in evaporator. Applied Thermal Engineering. 2014;**62**(1):176-186

[16] Ran T, Qingsong AN, Lin SHI, Huixing Z, Xiaoye D. Performance analyses of supercritical organic Rankine cycles (Orcs) with large variations of the thermophysical properties in the

pseudocritical region. 3rd International Seminar on ORC Power Systems, October 12-14, 2015, Brussels, Belgium. 2015:1-10. Paper ID: 53

[17] Moloney F, Almatrafi E, Goswami DY. Working fluid parametric analysis for regenerative supercritical organic Rankine cycles medium geothermal reservoir organic Rankine cycles for for medium geothermal reservoir temperatures. Energy Procedia. 2017;**129**:599-606

[18] Chowdhury JI, Nguyen KB, Thornhill D, Douglas R, Glover S. Modelling of organic Rankine cycle for waste heat recovery process in supercritical condition. International Journal of Mechanical, Aerospace, Industrial and Mechatronics Engineering. 2015;**9**(3):477-482

[19] Rentizelas A, Karellas S, Kakaras E, Tatsiopoulos I. Comparative technoeconomic analysis of ORC and gasification for bioenergy applications. Energy Conversion and Management. 2009;**50**(3):674-681

[20] Karytsas C, Mendrinos D, Radoglou G. The Current Geothermal Exploration and Development of the Geothermal Field of Milos Island in Greece GHC Bulletin. 2004;**25**:17-21

[21] Frick S, Kranz S, Saadat A, Huenges E. Design approach for geothermal binary power. Plants. 2009

[22] Müller-steinhagen H, Trieb F. Concentrating Solar Power—A Review of the Technology. Vol. 18. Stuttgart, Germany: Institute of Technical Thermodynamics, German Aerospace Centre; 2004. pp. 43-50

[23] Ford G. CSP: Bright future for linear fresnel technology? Renewable Energy Focus. 2008;**9**(5):48-49

[24] Canada S, Cohen G, Cable R, Brosseau D, Price H. Parabolic Trough Organic Rankine Cycle Power Plant. United States. 2005. p. 5

[25] Bundela PS, Chawla V. Sustainable development through waste heat recovery. American Journal of Environmental Sciences. 2010;**6**(1):83-89

[26] Engin T, Ari V. Energy auditing and recovery for dry type cement rotary kiln systems—A case study. Energy Conversion and Management. 2005;**46**(4):551-562

[27] Perullo CA, Mavris DN, Fonseca E. An Integrated Assessment of an Organic Rankine Cycle Concept for Use in Onboard Aircraft Power Generation. In ASME Turbo Expo: Turbine Technical Conference and Exposition American Society of Mechanical Engineers. 2013

[28] Le VL, Feidt M, Kheiri A, Pelloux-Prayer S. Performance optimization of low-temperature power generation by supercritical ORCs (organic Rankine cycles) using low GWP (global warming potential) working fluids. Energy. 2014;**67**:513-526

[29] Muhammad I, Park BS, Kim HJ, Lee DH, Muhammad U, Heo M. Thermo-economic optimization of Regenerative Organic Rankine Cycle for waste heat recovery applications. An International Journal Energy Conversion and Management. 2014;**87**(November):107-118

[30] Li L, Ge YT, Tassou SA. Experimental study on a small-scale R245fa organic Rankine cycle system for low-grade thermal energy recovery. Energy Procedia Energy Procedia 2017;**105**:1827-1832

[31] Roy J, Misra A. Parametric optimization and performance analysis of a regenerative organic Rankine cycle using R-123 for waste heat recovery. Energy. 2012;**39**(1):227-235

[32] Zhang C, Liu C, Xu X, Li Q, Wang S, Chen X. Effects of superheat and internal heat exchanger on thermo-economic performance of organic Rankine cycle based on fluid type and heat sources. Energy. 2018;**159**:482-495

[33] Brizard A, Dolz V, Galindo J, Royo-Pascaul L. Dynamic modeling of an organic Rankine cycle to recover waste heat for transportation vehicles. Energy. 2017;**129**:192-199

[34] Radulovic J, Benedikt K. Utilisation of diesel engine waste heat by Organic Rankine Cycle. Applied Thermal Engineering. 2015;**78**:437-448

[35] Feng Y, Zhang Y, Li B, Yang J, Shi Y. Comparison between regenerative organic Rankine cycle (RORC) and basic organic Rankine cycle (BORC) based on thermoeconomic multi-objective optimization considering exergy efficiency and levelized energy cost (LEC). Energy Conversion and Management. 2015;**96**:58-71

[36] Bianchi M et al. Experimental performance of a micro-ORC energy system for low grade heat recovery. Energy Procedia. 2017;**129**:899-906

[37] Ismail H, Aziz AA, Rasih RA, Jenal N, Michael Z, Roslan A. Performance of organic Rankine cycle using biomass as source of fuel. Journal of Advanced Research in Applied Sciences and Engineering Technology. 2016;**4**(1):29-46